Advanced Structured Materials

Volume 128

Series Editors

Andreas Öchsner, Faculty of Mechanical Engineering, Esslingen University of Applied Sciences, Esslingen, Germany

Lucas F. M. da Silva, Department of Mechanical Engineering, Faculty of Engineering, University of Porto, Porto, Portugal

Holm Altenbach⑩, Faculty of Mechanical Engineering,
Otto von Guericke University Magdeburg, Magdeburg, Sachsen-Anhalt, Germany

Common engineering materials reach in many applications their limits and new developments are required to fulfil increasing demands on engineering materials. The performance of materials can be increased by combining different materials to achieve better properties than a single constituent or by shaping the material or constituents in a specific structure. The interaction between material and structure may arise on different length scales, such as micro-, meso- or macroscale, and offers possible applications in quite diverse fields.

This book series addresses the fundamental relationship between materials and their structure on the overall properties (e.g. mechanical, thermal, chemical or magnetic etc.) and applications.

The topics of *Advanced Structured Materials* include but are not limited to

- classical fibre-reinforced composites (e.g. glass, carbon or Aramid reinforced plastics)
- metal matrix composites (MMCs)
- micro porous composites
- micro channel materials
- multilayered materials
- cellular materials (e.g., metallic or polymer foams, sponges, hollow sphere structures)
- porous materials
- truss structures
- nanocomposite materials
- biomaterials
- nanoporous metals
- concrete
- coated materials
- smart materials

Advanced Structured Materials is indexed in Google Scholar and Scopus.

More information about this series at http://www.springer.com/series/8611

Adel Mellit · Mohamed Benghanem
Editors

A Practical Guide
for Advanced Methods
in Solar Photovoltaic Systems

 Springer

Editors
Adel Mellit
Renewable Energy Laboratory
University of Jijel
Jijel, Algeria

Mohamed Benghanem
Faculty of Science, Physics Department
Islamic University
Madinah, Saudi Arabia

ISSN 1869-8433   ISSN 1869-8441   (electronic)
Advanced Structured Materials
ISBN 978-3-030-43472-4   ISBN 978-3-030-43473-1   (eBook)
https://doi.org/10.1007/978-3-030-43473-1

This Springer imprint is published by the registered company Springer Nature Switzerland AG
The registered company address is: Gewerbestrasse 11, 6330 Cham, Switzerland

Preface

The development in solar photovoltaic (PV) technology is growing very fast in recent years due to technological improvement, cost reductions in materials and government support for renewable energy-based electricity production. PV is playing an important role to utilize solar energy for electricity production worldwide. With reference to IRENA, the PV market is growing rapidly with worldwide around 24 GW in 2010 and also growing at an annual rate of 35–40%, which makes photovoltaic as one of the fastest growing industries. The total solar PV capacity almost growing the next ten years, as reported by the International Renewable Energy Agency (IRENA), from a global total of 480 GW in 2018 to 2840 GW by 2030, and to 8 519 GW by 2050. The same source mentioned that in the last decade, the globally installed capacity of off-grid solar PV has grown more than tenfold from roughly 0.25 GW in 2008, to almost 3 GW in 2018. Off-grid solar PV is a key technology for achieving full energy.

Solar panels have improved substantially in their efficiencies and power output over the last few decades. The efficiency of solar cell represents the most important parameter in order to establish this technology in the market. Due to the decrease cost of PV modules, millions of PV systems were installed around the world. To better exploit the produced power by these installations and keep them working with good reliability and safety, these installations need to be monitored and supervised periodically. Recently, many techniques have been developed to control, supervise, optimize, monitor and diagnose these kind of systems. They vary in terms of complexity algorithms, cost implementation, effectiveness and feasibility.

This book aims to offer the reader's background on solar cells development and the latest advanced methods in PV system applications. This book is mainly divided into two parts, the first one deals with theoretical and experimental in solar cells, including silicon, thin films, quantum and organic solar cells. The second part of this book provides some advanced methods in photovoltaic systems, including the application of artificial intelligence (AI) techniques in control, optimization, fault diagnosis and forecasting. The main goal of this book is to give postgraduate students and researchers a resource on the recent development in solar cells and photovoltaic applications. This book is also useful for engineers and industry personnel who want to have a thorough understanding of the subject.

This book is comprising of 13 chapters, six chapters cover subjects related to solar cell theory, modelling, simulation and design, while the rest seven chapters cover a variety of PV methods and applications, including, reconfiguration and optimisation of PV arrays, control and PV output power forecasting, PV integration in extra-high voltage level, energy storage, monitoring and fault diagnosis of PV plants.

Chapter 1 is dedicated to solar radiation and solar simulators, which are widely used by PV researches. In this chapter, the fundamentals of solar radiation and the components of solar radiation in the atmosphere are briefly presented. Besides, solar simulators used in PV tests are emphasized, classification of solar simulators, international standards and light sources are explained in detail. Also, as a case study, LED-based solar simulator is designed.

Chapter 2 is devoted to study the performance of solar cells using thermoelectric module (TEM) as cooling system. A hybrid PV/TEM system is proposed for PV applications in hot sites (specific climate: Sahara regions). The investigated solar cells are mono-crystalline silicon, but the method could be used for cooling other solar cell technologies such thin-film solar cells.

In Chap. 3, an optical optimization of the tandem structure composed of organic photovoltaic (OPV) cells based on interpenetrating blend materials P3HT:PCBM and pBBTDPP2:PCBM is presented. The optical optimization of OPV based on P3HT:PCBM interpenetrating blend is firstly performed to determine the optimal geometry of the stack giving the best optical properties. The developed MATLAB program is based on the transfer matrix formalism. Different simulation steps are provided and discussed in details.

Chapter 4 presents a theoretical study, and simulation of a compressively strained $GaAs_xP_{1-x}$ and tensile strain $GaN_yAs_xP_{1-x-y}$ quantum well active zones with the aim to be inserted in solar cells. The chapter presents and compares the ternary GaAsP/GaP and quaternary GaNAsP/GaP quantum well structures (QWs) by modelling these two types of systems.

Chapter 5 focuses on a comprehensive review on organic solar cells. The chapter provides different materials, devices structures and different processing techniques for the fabrication of OPV cells. The manufacturer of these types of solar cells uses new process to get best efficiencies with low cost by using printing techniques and photoactive layers based on polymer materials. Also, many scientific research works are presented, and some illustrations about processing techniques, such as roll-to-roll techniques, for the design of OPV cells are presented in this chapter.

Chapter 6 investigates in detail the effect of different dopants (Al, Sn and Cu) on the structure, texture and optical properties of ZnO thin films. Al-doped ZnO (AZO), Sn-doped ZnO (TZO) and Cu-doped ZnO (CZO) films are synthesized by chemical spray pyrolysis technique on glass substrates. The so-obtained films crystallized in hexagonal Wurtzite polycrystalline structure. The pole figures show that all the thin films have (0002) as the preferred orientation along the c-axis with the highest level was obtained in TZO. The morphology film was significantly affected by doping type. The transmittance spectra of all the films point out highly transparent in the visible range with an average transmittance higher than 80% for TZO and AZO films but with an average transmittance equal to about 70% for CZO film. The optical band gap values of the films are found to be 3.30 eV, 3.28 eV and 3.27 eV for Al-, Sn- and Cu-doped ZnO thin films, respectively. The Urbach energy of the films was also calculated in this chapter.

In Chap. 7, a detailed description of a new method for reconfiguring the dynamic PV array under repeating shade conditions is provided. The repeating shades are often caused in PV installations, especially in residential installations where PV modules can be subjected to shades occurred by nearby buildings or trees. The method is based on logic gates and aims to minimize the processing time in the way that controller does not have to perform exhaustive calculations at each shade condition to achieve the optimal configuration of the PV generator. Simulation of 2x2 size dynamic photovoltaic (DPV) array has been carried out. Experimental tests of 1x1 size DPV array under different solar irradiance conditions have been also conducted.

Chapter 8 provides a comprehensive description of dynamic batteries behaviour, encountered problems in the PV systems with solutions proposal in terms of modelling and control. The storage in renewable energy systems (RESs) especially in the PV stations is still a major issue related to their unpredictable and complex working. Due to the continuous changes of the source outputs, several problems can be encountered for the sake of modelling, monitoring, control and lifetime extending of the storage devices. Therefore, several storage devices were introduced in the practice such as: pumped hydro, compressed air, flywheel, supercapacitors and electrochemical storage. However, the electrochemical storage especially the storage by battery bank is still the most used in PV systems. According to the performances and the features needed in such systems, two battery types can be distinguished, namely lithium-ion and lead-acid-based batteries.

Chapter 9 focuses on the integration of renewable energy sources in the main grid, a case study in Istanbul, Turkey is presented. The share of variable renewable sources, especially solar energy, in total installed power capacity increases day by day. The power systems with the integration of solar energy sources transform. A power system to cope with high shares of variable solar generation needs to be flexible. In this chapter, the power system flexibility concept, the effect of variable renewable energy penetration, especially power plants on power systems flexibility, are examined. In addition, simulation studies are carried out for PV power systems penetration into extra-high-voltage levels (EHVL), necessary regulations for grid codes are determined and solution methods are presented.

The main goal of Chap. 10 is to show the set-up a well-defined method to identify and properly train the Hybrid Artificial Neural Network both in terms of a number of neurons, hidden layers and training set size in order to perform the day-ahead power production forecast applicable to any PV plant, accurately. Therefore, this chapter has been addressed to describe the adopted hybrid method (PHANN—Physical Hybrid Artificial Neural Network) combining both the deterministic Clear Sky Solar Radiation Algorithm (CSRM) and the stochastic Artificial Neural

Network (ANN) method in order to enhance the day-ahead power forecast. In this chapter, the main results obtained by applying the abovementioned procedure specifically referred to the available data of the PV power production of a single PV module are presented.

Chapter 11 deals with the control strategy of stand-alone hybrid photovoltaic/wind/battery power system. The goal is to examine the performance of the power system under several conditions of generation and demand. A centralized power management system is established to supervise the power flow between the generation units and user loads. Local controllers of the (photovoltaic/wind/battery) power sources are designed based on simple control schemes. The overall system is simulated in the MATLAB/Simulink environment using Xilinx System Generator (XSG) tool for possible implementation on field-programmable gate array (FPGA) board. The simulation results are provided in order to demonstrate the accuracy and feasibility of the designed control scheme.

Chapter 12 aims at investigating the performances of three different PV technologies: poly-crystalline silicon (Poly C-Si), copper indium gallium selenide (CIGS) and cadmium telluride (CdTe) in terms of several aspects. A simple PV model based on manufacture's datasheet has been used. Modelling and simulation I-V curves of different PV modules technology-based MATLAB-Simscape is described in details. A test facility is employed to carry out the required tests for assessing the proposed PV model. Obtained experimental results under different climate conditions are compared with simulated ones. The comparison is carried out by evaluating four statistical errors with a view of measuring the accuracy of the proposed model in predicting the I-V and P-V characteristics.

Chapter 13 presents a brief survey on the recent application of artificial intelligence (AI) techniques in fault diagnosis of PV plants. AI-based methods are mainly used to identify and classify the type of faults that can be happened in PV systems, particularly in DC side. The methods will be presented and discussed in terms of complexity implementation, fault identification, classification and localization. Localization of fault in large-scale PV plants remains a challenging issue, to date no AI-based method was applied to localize fault in large-scale PV plants. It is believed that this brief review can help users and researchers to get a clear idea on the potential application of AI techniques in this interesting field.

We believe that the readers will find this book as useful for solar cells and solar PV applications. Also, we hope that this book will contribute in the field of solar cells and PV system applications.

Jijel, Algeria   Adel Mellit
Madinah, Saudi Arabia   Mohamed Benghanem

Acknowledgements First, we would like to thanks Springer Publisher for accepting our book proposal. We would like also to express our deep gratitude to all contributors in this book.

Contents

Part I
SOLAR CELLS: Theory, Modeling and Simulation

Chapter 1
Solar Irradiation Fundamentals and Solar Simulators

V. Esen, Ş. Sağlam, and B. Oral

Abstract This chapter is prepared for introduction to solar radiation and solar simulators, which are widely used photovoltaic researches. In this study, the fundamentals of solar radiation and the components of solar radiation in the atmosphere are briefly expressed. Besides, solar simulators used in PV tests are emphasized, classification of solar simulators, international standards and light sources are explained in detail. Also, as a case study, it is designed LED-based solar simulator.

Keywords Solar irradiation · Solar simulator · Light emitting diodes · Photovoltaic · Spectroradiometers

1.1 Solar Radiation

Solar radiation can be defined as electromagnetic radiation emitted by the Sun in the spectrum ranging from X-rays to radio waves [1]. 99% of the energy of solar radiation is at the wavelength of 150–400 nm and includes the ultraviolet, visible and infrared regions of the solar spectrum. About 40% of the solar radiation reaching the earth's surface in the cloudless days is visible radiation in the wavelength range of 400–700 nm. 51% is infrared radiation between 700 and 4000 nm [2]. The Sun's rays are distributed by clouds and airborne particles, reflected and broken. This radiation is usually used in solar technology applications from 300 to 4000 nm wavelength. Solar radiation differs according to astronomical factors and weather conditions [3]. According to NASA's Solar Radiation and Climate Experiment (SORCE), the annual average value of Total Solar Irradiance (TSI) from the Sun to the Earth's atmosphere is 1360.8 ± 0.5 W/m^2 [4]. Figure 1.1 shows the distribution of direct solar radiation in the world according to Global Solar Atlas data [5].

V. Esen · Ş. Sağlam (✉) · B. Oral
İstanbul, Turkey
e-mail: ssaglam@marmara.edu.tr

© The Editor(s) (if applicable) and The Author(s), under exclusive license
to Springer Nature Switzerland AG 2020
A. Mellit and M. Benghanem (eds.), *A Practical Guide for Advanced Methods
in Solar Photovoltaic Systems*, Advanced Structured Materials 128,
https://doi.org/10.1007/978-3-030-43473-1_1

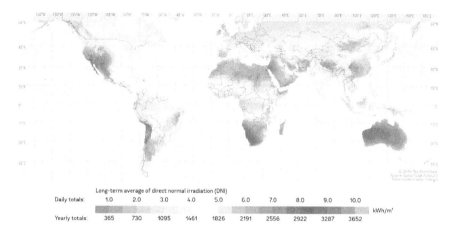

Fig. 1.1 Direct normal irradiation

Solar radiation reaches the surface of the Earth at a very long distance. For this reason, in the modeling of solar radiation, factors such as the Earth's atmosphere, surface and various objects in the world should be taken into account.

According to Harrouni [6], "When solar radiation enters the Earth's atmosphere, a part of the incident energy is removed by scattering or absorption by air molecules, clouds and particulate matter usually referred to as aerosols. The radiation that is not reflected or scattered and reaches the surface straightforwardly from the solar disk is called direct or beam radiation. The scattered radiation which reaches the ground is called diffuse radiation. Some of the radiation may reach a panel after reflection from the ground, and is called the ground reflected irradiation." Figure 1.2 illustrates the various components of solar radiation on intercepting surfaces.

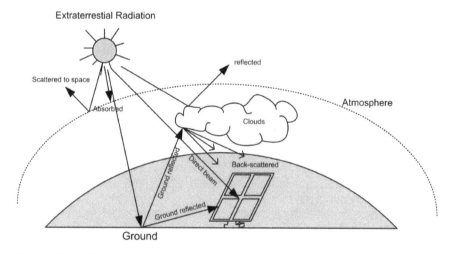

Fig. 1.2 Solar radiation components segregated by the atmosphere and surface

One of the benefits of solar radiation is electricity production. Electricity production via solar radiation is usually provided by photovoltaic devices. Photovoltaic devices are nonlinear energy sources with durable and simple designs that require little maintenance and convert solar radiation directly into electrical energy. It is very important that the characteristic values of photovoltaic devices such as current–voltage curve, short-circuit current, open-circuit voltage and maximum power are determined under real atmospheric conditions [7–10]. However, real atmospheric conditions are not preferable due to factors such as the intensity and the spectral distribution of solar radiation, geographical location, time, day of the year, climate conditions, the composition of the atmosphere, variation in altitude and weather conditions [11]. Instead, solar simulators are preferred for reasons such as simplicity, reproducibility and reliability [12]. For this reason, solar simulators are an integral part of current–voltage (I–V) characterization. This is because the I–V measurement requires a calibrated source that corresponds to real daylight and conditions that can be changed on demand to illuminate a photovoltaic panel.

1.2 Solar Simulators

In order to evaluate their performances, photovoltaic devices are rated under the so-called Standard Test Conditions, corresponding to an irradiance of 1000 W/m^2, an AM (air mass) 1.5 spectrum and a device temperature of 25 °C [13]. I–V measurement is carried out under natural sunlight in the outdoor environment or in a closed laboratory environment with the help of a solar simulator [12]. Solar simulators are tools that provide spectral and optical composition similar to sunlight intensity. The fundamental aim of these tools is to test solar cells and photovoltaic panels under controlled laboratory conditions [14–17]. In today's world, as the usage of renewable energy resources has been increasing it is important for both photovoltaic tool producers and consumers that tests are conducted, due to photovoltaic tools having low efficiency [18]. A solar simulator mainly consists of three parts; light and power sources, an optical filter to change beam properties in order to fulfill requirements, and control elements to operate the simulator [7, 19]. Carbon arc lamps, sodium vapor lamps, argon arc lamps, quartz tungsten halogen lamps, mercury xenon lamps, xenon arc, xenon flash lamps, metal halide lamps, LED and supercontinuum laser light sources are investigated within the scope of the present study [15, 20–25]. Xenon arc lamps are the most commonly used light sources among conventional solar simulators [26, 27]. Since there are intensity and spectral component differences between natural sunlight and artificial light, xenon arc lamps are modified using filters to obtain the natural sunlight spectrum [28]. Test standards for the terrestrial application of photovoltaic panels have been presented in the research conducted by ERDA and NASA. A report published after the studies conducted in 1975 and 1977, provided a detailed explanation of solar simulators as well as the standard procedures for terrestrial photovoltaic measurement [29]. In this report, the intensity was selected as 1000 W/m^2, AM 1.5 spectral component and 25 °C ambient temperature was chosen as an STC, however, in today's commercial solar simulators both of these are used as the ASTM (American Society for Testing and Materials) standards [30, 31]. Solar simulators mainly simulate natural sunlight in two categories: space radiation and terrestrial radiation. The ASTM 927-10 [32] JIS (Japanese Industrial Standard) C 8912 [33] and IEC (International Electrotechnical Commission) 60904-9 standards [34] for the solar simulation of terrestrial photovoltaic tests are accepted to evaluate three criteria: spectral distribution of solar simulator performance, spatial difference and temporal constancy [35, 36]. The number of solar beams that fall on the Earth's surface depends on factors such as latitude, longitude, time of day and time of year [32]. According to ASTM G173-03 standard shows, the spectral distribution of the Sun on the Earth's surface (extraterrestrial spectral irradiance, direct normal spectral irradiance and the global total spectral irradiance) is shown in Fig. 1.3.

In the application, the air mass for the photovoltaic panel test was standardized as AM 0 (the Sun's radiation in space), AM 1 D (Direct), AM 1 G (Global), AM 1.5 D, AM 1.5 G, AM 2 D and AM 2 G [37, 38]. According to Riordan and Hulstron; air mass refers to the relative path length of the direct solar beam through the atmosphere. When the Sun is directly overhead (at zenith), the path length is 1.0 (AM 1.0). AM 1.0 is not synonymous with solar noon because the Sun is usually not directly overhead at solar noon in most seasons and locations. When the angle of the Sun from zenith (i.e., the zenith angle, θ) increases, the air mass increases (approximately by $\sec\theta$) so that at about 48° from the vertical the air mass is 1.5 and at 60° the air mass is 2.0 [39].

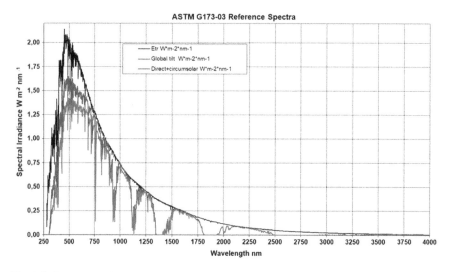

Fig. 1.3 Spectral distribution of the Sun on the Earth's surface

1.2.1 Solar Simulator Standards

International standard references for solar simulator illuminations are stated below:

- IEC 60904-9 Solar Simulator Performance Requirements [34].
- ASTM G173-03 Standard Tables for Reference Solar Spectral Irradiances: Direct Normal and Hemispherical on 37° Tilted Surface [32].
- ASTM E927-10 Standard Specification for Solar Simulation for Photovoltaic Testing [40].
- ASTM E948-16 Standard Test Method for Electrical Performance of Photovoltaic Cells Using Reference Cells Under Simulated Sunlight [41].
- ASTM E973-16 Standard Test Method for Determination of the Spectral Mismatch Parameter between a Photovoltaic Device and a Photovoltaic Reference Cell [42].
- JIS C 8912 Solar simulators for crystalline solar cells and modules [33].

1.2.2 Solar Simulator Classes for Photovoltaic Devices

According to ASTM E927 (Standard Specifications for Solar Simulation for Terrestrial Photovoltaic Testing) and IEC 60904-9, simulation performances of solar simulators are defined under three classes: Class A, Class B and Class C [43]. This classification determines three main criteria: spectral match, spatial non-uniformity of irradiance and temporal instability [15]. These criteria are shown in Table 1.1 [17].

Table 1.1 Classifications of simulator performance

Performance parameters	ASTM	IEC
Spectral match		
Class A	0.75–1.25	0.75–1.25
Class B	0.6–1.4	0.6–1.4
Class C	0.4–2.0	0.4–2.0
Spatial non-uniformity		
Class A	$\leq 3\%$	$\leq 2\%$
Class B	$\leq 5\%$	$\leq 5\%$
Class C	$\leq 10\%$	$\leq 10\%$
Temporal instability		
Class A	$\leq 2\%$	$\leq 2\%$
Class B	$\leq 5\%$	$\leq 5\%$
Class C	$\leq 10\%$	$\leq 10\%$

Under these criteria, the highest class is stated as Class A and the lowest class is stated as Class C [15]. The spectral match is important to provide one to one correspondence of real-world situations and test conditions [43]. According to Mohan's study [23], spectral match (SM) is calculated as the ratio of the actual percentage of irradiance falling on the interval of concern and the required percentage of irradiance. This calculation is shown in Eq. (1.1).

$$SM = \frac{\textbf{Actual irradiance in the interval}}{\textbf{Required irradiance in the interval}} \qquad (1.1)$$

The "actual irradiance in the interval" required to find the spectral match is found by Eq. 1.2:

$$\textbf{Act. irradiance in the interval} = \frac{\int_{\lambda_n}^{\lambda_{n+1}} S(\lambda)d\lambda}{\int_{400}^{1100} S(\lambda)d\lambda} \qquad (1.2)$$

$S(\lambda)$ in Eq. 1.2 is the light source's spectral irradiation. Spectral irradiation is the distribution of light as a function of wavelength. λ_n is the starting point of the corresponding wavelength range, and λ_{n+1} is the ending point. The spectral matching is directly related to the light source used in the solar simulator. If the source's spectral composition source matches the spectral glow of the AM 1.5 G reference spectral glow, then the spectrum will be an exact match. The ability to simulate the expected spectral coherence will be of great help in designing a solar simulator. Spatial non-uniformity (SNU) is the hardest property that minimizes hot spots due to condensation of light in certain points, especially in simulators with large surface areas, affecting cell performance tests and the need for repetition. SNU is calculated in which E_{max} is the maximum intensity on the given test section and E_{min} is the minimum intensity on the given Eq. 1.3 [44].

$$SNU = \frac{E_{max} - E_{min}}{E_{max} + E_{min}} \times \%100 \qquad (1.3)$$

The third criterion used in classifying simulators is temporal instability. It explained this concept as follows: "The temporal instability of irradiance is calculated in a manner similar to SNU but with E measured at a particular point on the test plane during the time interval of data acquisition" [44].

In Table 1.2, different AM factor classifications are provided for wavelength intervals [44].

Table 1.2 Distribution of irradiance performance requirements

Wavelength (nm)	Percentage of total irradiance
	AM 1.5 G (%)
400–500	18.4
500–600	19.9
600–700	18.4
700–800	14.9
800–900	12.5
900–1100	15.9

1.3 Light Sources of Solar Simulators for Photovoltaic Devices

Light source selection is the most important part of solar simulator design for the simulation of sunlight and its intensity, and spectral properties of light source, illumination pattern, collimation, light flow stability and light range should be taken into account for the selection [45]. In solar simulator applications, different light sources are used to simulate sunlight according to the ASTM standards [7]. The most commonly used light sources are short arc and long arc xenon lamps. In some simulator designs, metal halide arc lamps, carbon arc lights and quartz tungsten halogen lamps are selected as light sources [38]. Light emitting diodes (LEDs) are more often preferred as light sources in research compared to the traditional light sources used today since they are more advantageous in terms of cost, compactness and power consumption [37].

Classification and the years they started to be used light sources of solar simulators are shown in Table 1.3.

Table 1.3 Wavelength range and historical development of solar simulator light sources

Light sources	Years	Wavelength range (nm)
Carbon arc	1960	350–700
Xenon arc—mercury xenon	1961	185–2600
Quartz tungsten halogen	1962	250–2500
Argon arc	1972	275–1525
Multi light source	1990	185–2600
LED	2003	350–1100
Metal halide	2005	200–2600
Super continuum laser	2011	480–900

1.3.1 LED (Light Emitting Diode)

Compared to the traditional light sources used in research today, LEDs are being used as light sources since LEDs have advantages such as low cost, compactness and lower power consumption [46]. LED light source solar simulators can simulate the AM 1.5 spectrum with a Class A spectral fit at a wavelength range of 350–1100 nm. LED solar simulators deliver high performance in power consumption in steady and pulsed mode [47]. LED solar simulators have a high-quality character as well as very different optical characteristics in comparison with standard light sources [48, 49]. Another advantage of LED solar simulators is that they can be accurately controlled and offset the output density in less than a millisecond (typically a few microseconds). The possibility of stabilizing the operating parameters of the LED opens up new ways of determining the short- and long-term effects on the solar cells tested using the same solar simulator [50, 51].

LEDs were first used as light sources in solar simulations in the studies by Kohraku and Kurokawa [52]. They worked on new measurement methods in solar cells with LED solar simulators and tried to estimate spectral reactions with different and multiple LEDs. In another study conducted by them [18]. Four-color LEDs were used for solar cell measurements in the solar simulator they designed, where they stated that $I–V$ characteristic measurements of solar cells were significant under the AM 1.5 spectrum and LED simulators could be developed and used in the future applications [53]. It investigated the advantages of LED usage in the characteristic measurements of photovoltaic solar panels, and they stated that LEDs can control the spectral values that are compatible with the AM 1.5 standards within microseconds [54]. They are cost-efficient, their life cycle is long and calibration is easy, and thus, they are strong candidates for light sources in solar simulators in future applications. Tsuno et al. brought a new approach to the capabilities of LEDs in solar simulators and showed that in comparison with conventional light sources, LEDs are more efficient, they are brighter light sources, they have a long life cycle and their costs are decreasing every day [54]. Jang and Shin worked on changing spectrum and flash speed LED-based photovoltaic measurement systems and established that LED simulator in stationary situations can conduct flexible controls on multiple flash tests they stated LED solar simulators are valid for solar cell characterization and achieved LED solar simulator thermal optimization [46]. Krebs et al. established an LED-based solar simulator application that can calibrate itself with lamps that have adjustable geometric shapes [55]. Kolberg et al. conducted a study on matching expanded ultraviolet (UV) and infrared (IR) spectral values with the AM 1.5 standards in LED solar simulators [56]. Meng et al. worked on the high values of LED solar simulator radiation characteristics and optimization, and they gained tangible results in hexagonally designed light source placement [32]. Kolberg et al. [57] improved the previous work, increased LED effectiveness in solar simulators and obtained measurements close to real sunlight values [57]. In this study, spectral values of LED-based solar simulators were adjusted, and values close to the solar spectrum were obtained. Plyta et al. investigated the LED solar simulator potential and first stated that LED light source had strong optic light collection efficiency and took the collimation to the highest level; and secondly, LED wavelengths were appropriate for an Class A simulator and the distribution of light was adequately homogenous [58]. Linden et al. developed an LED solar simulator with an adjustable spectrum, which was easy to produce and had a modular design [59]. In this simulator, single and multiple solar cell structures were used. This simulator showed A+ performance with illumination integrity and spectral properties. Novickovas et al. designed a solar simulator consisting of 19 high-power compact LEDs inside an AAA class test area of 5 cm [28]. Luka et al. applied external quantum efficiency (EQE) method to LED solar simulators and showed that it was an alternative for standard measurements [60]. In this study, half-second measuring time, global search and measurement without any additional EQE testing tools show advantages. Leary et al. used Oriel-VeraSol-LED and xenon light source Oriel-Sol3A solar simulator of Newport's LED light source to demonstrate the $I–V$ characteristics of the photovoltaic device, thus indicating that there is no difference in the $I–V$ response of the photovoltaic devices [61]. They show that the LED solar simulator provides the same performance with that of a xenon light source solar simulator at 400–1100 nm wavelength.

1.3.2 Supercontinuum Laser

Today, the light sources in solar simulators are typically xenon arc lamps and LEDs. Accordingly, lamps and LEDs have a big and constant radiation spectrum and it is hard to focus on the efficient spectral area, and thus, these light sources are hard to apply spectrally into efficient measurement systems [34]. For example, LEDs can radiate to a larger angled area compared to supercontinuum lasers. However, they are insufficient in UV and IR wavelengths of the solar spectrum [62]. The supercontinuum laser is a high-power, a broadband light source that provides greatly improved optical compatibility for photovoltaic materials and devices [63]. Therefore, lately, high-power supercontinuum lasers can be considered as commercial products since they can be used as a light source ranging from visible to infrared spectrums [64]. Lasers are powerful and easy to concentrate but have unrealistically narrow spectra [65]. Relatively higher-power supercontinuum lasers have the potential to be used in photovoltaic measurements more commonly [66]. In recent studies where supercontinuum lasers were used as the light source in small regions, results were hopeful and it is predicted that this technology will further develop.

1.4 LED-Based Solar Simulator

The aim of LED-based solar simulator project was to design an efficient and affordable Class AAA solar simulator. After comparing with all other potential light sources used in solar simulators, it was determined that LEDs are the most convenient light sources. LEDs offer benefits such as low cost, long life, low energy consumption, compactness and adjustable spectrum. Compared to conventional light sources, LEDs are quite advantageous due to the above specifications. Furthermore, the spectra of the LEDs are controllable because they are driven by microprocessors. The performance criteria of a solar simulator are clearly mentioned in the IEC and ASTM927 standards. PV panel tests should be done with a solar simulator that has a 1000 W/m^2 irradiation intensity at 25 °C. One of the advantages of using LEDs as a light source is that performance criteria are met without the need for too many different colored LEDs.

This study revealed the following:

- LEDs are a convenient and efficient light source for solar simulators designed to be used in PV panel tests.
- LEDs reduce the cost of simulator production.
- Spectrum checking can be done easily with LEDs in solar simulators.
- LEDs are preferred in solar simulators thanks to their low energy consumption, lack of harmful effects of heating on PV panels and long life.
- A 1000 W/m^2 irradiation intensity, which is indicated in IEC and ASTM, can be obtained in solar simulators using LEDs.
- Spectral match, spatial non-uniformity and temporal instability, which are indicated in the IEC and ASTM standards, can be obtained at the Class A level in the solar simulators using LEDs.

1.4.1 LED-Based Solar Simulator Development Project

Before working on the development of a solar simulator, a related literature review was completed and commercial solar simulators in the market in which different light sources were used were investigated. As a result of this investigation, LEDs were determined according to the performance criteria mentioned in the IEC and ASTM standards. After the measurements, a convenient LED settlement has been done.

Spectral match criteria were obtained on the Class A level thanks to the LEDs chosen in the first step. Light intensity remained at the level of 350 W/m^2. To obtain a 1000 W/m^2 value, COB LEDs were used instead of cold and hot white power LEDs. The wavelengths of the LEDs and type details are shown in Table 1.4. The wavelengths of the cool white and warm white LEDs are shown in Fig. 1.4 [67]. The neutral white shown in the figure was not used in the study.

With the last combination of LEDs, a Class AAA solar simulator was produced with power and COB LEDs of six different wavelengths according to the IEC 60904-9 and ASTM E927 standards. For this purpose, wavelengths were measured in a mirror-covered box with power LEDs by means of a spectroradiometer that could also measure ultraviolet (UV) and infrared (IR) wavelengths. At the end of the study, spectral match (one of the requirements for solar simulators specified in the IEC

Table 1.4 Information on the power and COB LEDs used in the solar simulator

LED type	Wavelength, peak-centroid	LED part number
Blue	465–470 nm	OSRAM OSLON LB CP7P
Red	730–721 nm	OSRAM OSLON GF CS8PM1.24
Infrared 1	860–850 nm	OSRAM OSLON SFH 4715S
Infrared 2	950–940 nm	OSRAM OSLON SFH 4725S
Cool white	In Fig. 1	SEOUL SDW04F1CJ2C2E-V0
Warm white	In Fig. 1	SEOUL SDW84F1CJ1G10E-V0

Fig. 1.4 Cool white and warm white LED wavelengths

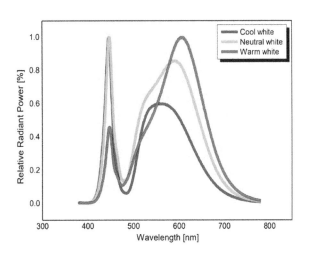

60904-9 and ASTM E927 standards) with power LEDs and six different wavelengths was obtained at the Class A level. At the same time, spatial non-uniformity of irradiance and temporal instability values were obtained at the Class A level. The workflow diagram which was monitored throughout the process from the beginning of the study is shown in Fig. 1.5.

1.4.2   Main Components of LED Solar Simulator

The main components of the solar simulator designed for this study are as follows:

- A box covered with an inner mirror
- An aluminum heat sink
- Cooling fans
- A power supply
- LED drivers
- Arduino Mega 2560

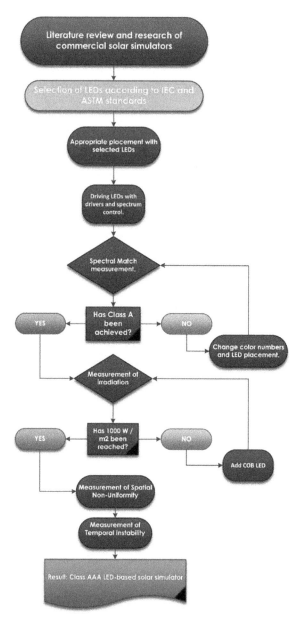

Fig. 1.5 Solar simulator design process diagram

- 5500 K cool white COB LEDs
- 3000 K warm white COB LEDs
- 470 nm blue LEDs
- 721 nm red LEDs
- 850 nm infrared LEDs
- 940 nm infrared LEDs.

A solar simulator was designed and manufactured using these materials. Figure 1.6 details all components of the solar simulator.

Fig. 1.6 LED solar simulator main components

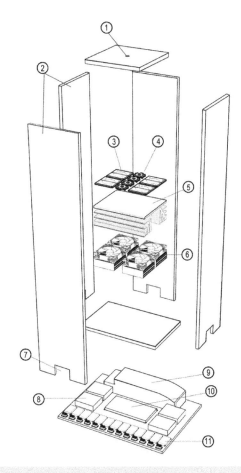

1- FIBER OPTIC CABLE HOLE, 2- BOX (with inner mirror),
3- COB LEDs, 4- POWER LEDs, 5- HEAT SINK,
6- COOLING FAN, 7- CABLE AND AIR HOLE,
8- COB LED DRIVERS, 9- POWER SUPPY,
10- ARDUINO MEGA, 11- LED DRIVERS

A total of 12 power LEDs in six different wavelengths were initially mounted with high sensitivity on star PCBs, and COB LEDs were mounted on the aluminum heat sink directly. The LEDs were placed symmetrically on the aluminum heat sink. The heat transfer for the assembly process was high, and the insulating tape was used.

Four cooling fans were added under the aluminum heat sink and placed inside the main box. The LEDs, heat sink and cooling fans were placed together, and the inner side of the box was covered with a mirror to allow for healthier wavelength measurement.

Appropriate LED drivers were chosen by considering the power and currents of the LEDs used. The LED drivers were regulated by an Arduino Mega 2560 microprocessor and software so that the desired spectral values could be reached. The microprocessor, LED drivers and power supply together formed a single power and control circuit in the lower part of the main box.

1.5   System Test Results and Discussion

In this section, the test results of the LED light source solar simulator prototype system will be given and examined. All calculations will be made using the formulas of the standards given in Sect. 1.2.

1.5.1   Spectral Match

The IEC 60904-9 and ASTM E927 standards mandate the test described in the previous sections. For this reason, measurements were made with a Spectral Evolution SR-500 Spectroradiometer. The fiber optic cable connected to the spectroradiometer was fixed to the other core so that the LEDs outside the ceiling center of the main box containing the LEDs could be viewed at a 90° right angle. The test setup consisted of the solar simulator, an SR-500 Spectroradiometer, a fiber optic cable and a laptop computer running DARWin SP software. Using the measurements, the LEDs were driven with the drive circuit, and the graphs and values in Fig. 1.7 are obtained. According to the measurements, each LED has its own characteristic wavelength. Warm and cool white COB LEDs and blue LEDs had wavelengths between 400 and 700 nm. The 700–850 nm chart was obtained with red 850 and 940 nm IR LEDs.

The comparison of the spectral match AM 1.5 G measured and calculated by the spectroradiometer is given in Fig. 1.8.

Thus, the values obtained for all wavelength ranges were at the Class A level, according to Table 1.5.

In Fig. 1.9, the orange lines show the range of Class A, and the dots show the range of the measured solar simulator values. The spectral match value in the wavelength range of 900–1100 nm was found to be at the boundary, despite the Class A level. In

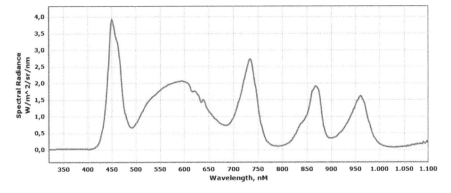

Fig. 1.7 Relative spectral irradiation measurements of the solar simulator design

Switch Solar Standard   Log to file

Spectrum | SR-500_SN1675028_00001

Wvl. Range	Class	AM1.5 Global Tilt	Irrad. %
400-500	A	18.4	19.0
500-600	A	19.9	23.2
600-700	A	18.4	17.5
700-800	A	14.9	15.6
800-900	A	12.5	11.6
900-1100	A	15.9	13.2

Fig. 1.8 Spectral match comparison (screenshot) for LED-based solar simulator design (using AM 1.5 G tilt)

future work, it will be possible to obtain a more normal range by making appropriate changes to the number of red LEDs and their combinations.

1.5.2 Spatial Non-uniformity

According to Grandi et al., "the irradiation produced by the simulator should have as low as possible non-uniformity. This parameter is definitely the most difficult to satisfy because solar radiation is very uniform. Uniformity is a measure of how the irradiance varies over the designed test area" [68]. Spatial non-uniformity is calculated by Eq. 1.3.

According to the IEC 60904-9 standards, non-uniformity on the surface of the solar simulator to be used in PV modules depends on the test room or the reflecting conditions in the test box. That is why it cannot be generalized, and uniformity

Table 1.5 Classifications of solar simulator spectral match performance

Spectral match	According to ASTM	According to IEC	Values in solar simulator						
			400–500 nm	500–600 nm	600–700 nm	700–800 nm	800–900 nm	900–1100 nm
Class A	0.75–1.25	0.75–1.25	1.03	1.16	0.95	1.04	0.92	0.83

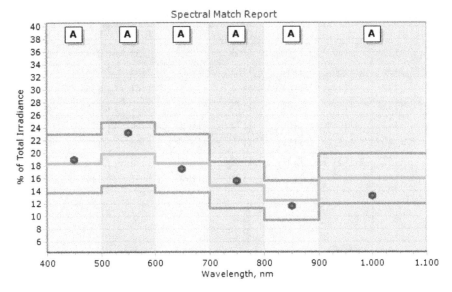

Fig. 1.9 IEC 60904-9 and ASTM E927 spectral match graph for the LED-based solar simulator design (using AM 1.5 G tilt)

is not assessed for each system. The determined test field is to be divided into at least 64 equal test positions (according to the field). The maximum homogeneity detector size will be the minimum value of the determined field divided into 64 pcs. The field that detector measurements cover is supposed to compose 100% of the determined test field. Measurement positions should be equally dispersed on the determined test field. In this application, testing equipment was created according to the IEC 60904-9 standards, and the test field was divided into 64 equal pcs. To obtain this, a measuring device was designed and mounted to a robotic device with $X-Y$ coordinates. Measurement intervals were determined by means of a computer software program.

This test was realized under two different distances to understand the homogeneity and difference between irradiation density. The distance between the measurement field and measuring device was 150 mm for the first measurement and 200 mm for the second one. The measurement for which the distance was 150 mm showed an irradiation value of 1173 W/m^2 on one of the 64 equal cells, D4. The minimum irradiation value was 1130 W/m^2 on the A8 cell. Detailed values are shown in Table 1.6. Figure 1.10 shows the non-uniformity surface graphics, which were measured from a 150 mm distance.

The irradiation value that resulted from the measurement from 200 mm was again 1027 W/m^2 in the D4 cell, and the minimum irradiation value was 995 W/m^2 in the A8 cell. Irradiation intensities in all cells are shown in Table 1.7.

Non-uniformity surface graphics measured from 200 mm are shown in Fig. 1.11. It was observed that irradiation intensity measured from 150 mm was more than

Table 1.6 Irradiance density (W/m^2) measurement values in 64 cells from 150 mm

	A	B	C	D	E	F	G	H
1	1137	1147	1156	1155	1156	1153	1138	1132
2	1137	1150	1162	1163	1165	1155	1148	1135
3	1138	1149	1163	1164	1165	1161	1154	1148
4	1138	1151	1164	1173	1169	1165	1154	1148
5	1142	1151	1162	1172	1168	1166	1156	1150
6	1140	1151	1158	1170	1171	1163	1155	1150
7	1142	1151	1154	1165	1164	1158	1152	1149
8	1130	1141	1151	1157	1159	1151	1143	1134

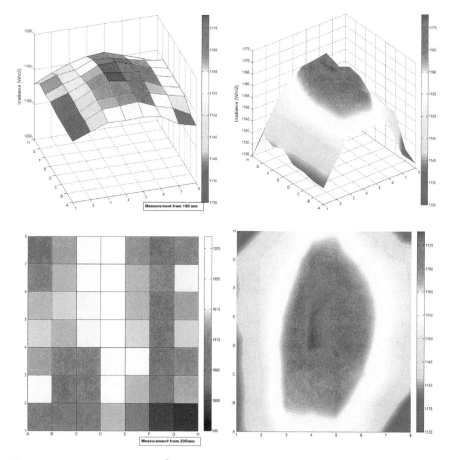

Fig. 1.10 Irradiance density (W/m^2) measurement for spatial non-uniformity from 150 mm

Table 1.7 Irradiance density (W/m²) measurement values in 64 cells from 200 mm

	A	B	C	D	E	F	G	H
1	1005	1004	1005	1013	1003	999	998	997
2	1017	1008	1009	1019	1013	1006	1012	1011
3	1012	1010	1010	1020	1016	1006	1010	1012
4	1014	1013	1015	1027	1014	1008	1014	1008
5	1013	1014	1015	1026	1013	1007	1012	1008
6	1010	1016	1016	1026	1013	1010	1015	1009
7	1007	1012	1017	1019	1011	1005	1010	1011
8	995	997	1000	1005	997	1000	1003	1005

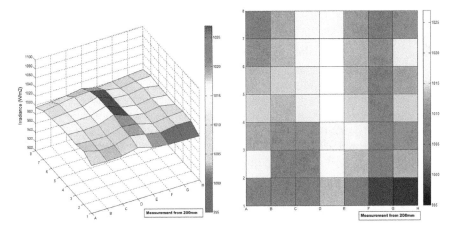

Fig. 1.11 Irradiance density (W/m²) measurement for spatial non-uniformity from 200 mm

that measured from 200 mm, while the homogeneity was slightly less successful compared with the measurement from 200 mm.

The spatial non-uniformity value of the measurement from 150 mm was 1.867%, and according to Table 1.8, it was a Class A value. It was noticed that the radiation intensity of the measurement from 200 mm was lower, while the homogeneity was more successful than the measurement from 150 mm. The spatial non-uniformity

Table 1.8 Classifications of solar simulator spatial non-uniformity performance

Spatial non-uniformity	According to ASTM (%)	According to IEC (%)	Distance (mm)	Calculation temporal instability (%)	Minimum measurement irradiance (W/m²)	Maximum measurement irradiance (W/m²)
Class A	≤3	≤2	150	1.867	1130	1173
			200	1.582	995	1027

value of the measurement from 200 mm was 1.582%, and according to Table 1.8, it was a Class A value.

1.5.3 Temporal Instability

The third solar simulator performance criterion is temporal instability. The formula shown in Equation 1.3 is used to calculate this criterion. However, E_{max} and E_{min} values are measured periodically from a point [44].

According to IEC 60904-9 standards, both long-term temporal instability and short-term temporal stability are to be assessed, and both assessments were done in this study.

1.5.3.1 Long-Term Instability Test

At first, a long-term test method was applied in this study. The solar simulator was turned on for 10 min without stopping. During its function, irradiation intensity was measured each minute from the beginning. The first measured irradiation value was 1024 W/m^2, whereas it was 1006 W/m^2 at the 10th minute. According to this value, the long-term temporal instability value was calculated as 0.91%. This value equals to Class A according to Table 1.9. An illustration of the results is provided in Fig. 1.12.

Table 1.9 Long-term instability calculation results

Temporal instability	According to ASTM (%)	According to IEC (%)	Calculation temporal instability (%)	Minimum measurement irradiance (W/m^2)	Maximum measurement irradiance (W/m^2)
Class A	≤2	≤2	0.91	1006	1024

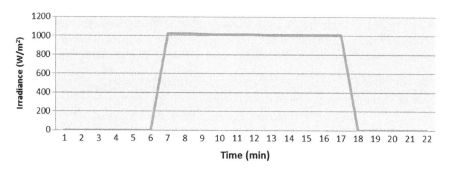

Fig. 1.12 Long-term temporal instability testing results

Table 1.10 Short-term instability calculation results

Temporal instability	According to ASTM (%)	According to IEC (%)	Pulse	Calculation temporal instability (%)	Minimum measurement irradiance (W/m²)	Maximum measurement irradiance (W/m²)
Class A	≤2	≤2	1	0.58	1028	1040
			2	0.58	1022	1034
			3	0.59	1020	1031

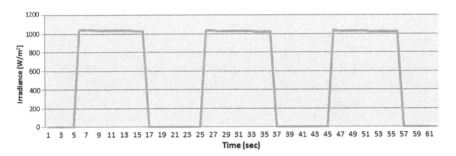

Fig. 1.13 Short-term temporal instability testing results

1.5.3.2   Short-Term Instability Test

After the long-term measurement, a short-term temporal instability test was conducted. The solar simulator was activated with a 10-s pulse. Values are shown in Table 1.10. According to these measurements, the short-term temporal instability values obtained are as follows: 0.58, 0.58 and 0.59%. These values are Class A according to Table 1.10. An illustration of the results is provided in Fig. 1.13.

1.6   Conclusion

Photovoltaic devices usually provide electricity production via solar radiation. They are nonlinear energy sources with durable and simple designs that require little maintenance and convert solar radiation directly into electrical energy. It is very important that the characteristic values of photovoltaic devices such as the current–voltage curve, short-circuit current, open-circuit voltage and maximum power are determined under real atmospheric conditions. For these reasons, it requires a calibrated source that corresponds to modifiable conditions to simulate real daylight.

In this section, international standards and light sources of solar simulators, which are widely used in cell and panel tests in photovoltaic researches are emphasized. In this context, an application of the simulator with LED technology used as a light source is presented. The offered is an efficient and cost-effective Class AAA solar simulator design.

Comparing light sources with all other potential sources used in solar simulators, it was determined that LEDs are the most convenient light sources. LEDs offer benefits such as low cost, long life, low energy consumption, compactness and adjustable spectrum. Compared to conventional light sources, LEDs are quite advantageous due to the above specifications. Furthermore, the spectra of the LEDs are controllable because they are driven by microprocessors. The performance criteria of a solar simulator are clearly mentioned in the IEC and ASTM927 standards. PV panel tests should be done with a solar simulator that has a 1000 W/m^2 irradiation intensity at 25 °C.

One of the advantages of using LEDs as a light source is that performance criteria are met without the need for too many different colored LEDs but only six different wavelengths. COB LEDs were preferred for cool and warm white wavelengths, and power LEDs were preferred for blue (470 nm), red (721 nm) and IR (850 and 940 nm) wavelengths. The spectral match values obtained by the spectroradiometer were as follows: 1.03 for 400–500 nm, 1.16 for 500–600 nm, 0.95 for 600–700 nm, 1.04 for 700–800 nm, 0.92 for 800–900 nm and 0.83 for 900–1100 nm (according to Eq. 1.1 and Table 1.2).

Therefore, solar simulators designed for the PV test can obtain Class A spectral matching using LEDs of six wavelengths. Spatial non-uniformity, which is one of the other two criteria, is at the Class A level in the solar simulator designed. A 1.867% value of spatial non-uniformity was obtained for the 150 mm measurements and a 1.582% value from 200-mm measurements. According to Table 1.8, 2% and lower values are considered Class A. Temporal instability, which is the last criterion, was measured as long term and short term according to the IEC standards. Thus, it is obvious that the values obtained from this test were lower than 2% and are considered Class A.

This study revealed the following:

- LEDs are a convenient and efficient light source for solar simulators designed to be used in PV panel tests.
- LEDs reduce the cost of simulator production.
- Spectrum checking can be done easily with LEDs in solar simulators.
- LEDs are preferred in solar simulators thanks to their low energy consumption, lack of harmful effects of heating on PV panels and long life.
- A 1000 W/m^2 irradiation intensity, which is indicated in IEC and ASTM, can be obtained in solar simulators using LEDs.
- Spectral match, spatial non-uniformity and temporal instability, which are indicated in the IEC and ASTM standards, can be obtained at the Class A level in the solar simulators using LEDs.

Developing LED technology will enable the realization of different studies in the future.

In future studies,

- Solar simulators requiring fewer LEDs that are more powerful can be produced.
- Spectrum controlling of the LEDs can be done more easily by developing driver technologies.
- LEDs or LED sets that can produce different wavelengths at the same time can be produced specifically for solar simulators.

References

1. Gueymard, C.A., Myers, D.R.: Solar radiation measurement: progress in radiometry for improved modeling. In: Badescu, V. (ed.) Modeling Solar Radiation at the Earth's Surface: Recent Advances, pp. 1–27. Springer, Berlin (2008)
2. Bhatia, S.C.: Advanced Renewable Energy Systems (Part 1 and 2). WPI Publishing (2014)
3. Şahin, A.D., Şen, Z.: Solar irradiation estimation methods from sunshine and cloud cover data. In: Badescu, V. (ed.) Modeling Solar Radiation at the Earth's Surface: Recent Advances, pp. 145–173. Springer, Berlin (2008)
4. Duffie, J.A., Beckman, W.A.: Solar Engineering of Thermal Processes. Wiley, Hoboken (2013)
5. World Bank Group: Direct normal irradiation map. In: Global Solar Atlas. https://globalsolaratlas.info/downloads/world (2019)
6. Harrouni, S.: Fractal classification of typical meteorological days from global solar irradiance: application to five sites of different climates. In: Modeling Solar Radiation at the Earth's Surface, pp. 29–54. Springer, Berlin (2008)
7. Li, D.H.W., Cheung, G.H.W., Lam, J.C.: Analysis of the operational performance and efficiency characteristic for photovoltaic system in Hong Kong. Energy Convers. Manag. **46**, 1107–1118 (2005)
8. Gxasheka, A.R., van Dyk, E.E., Meyer, E.L.: Evaluation of performance parameters of PV modules deployed outdoors. Renew. Energy **30**, 611–620 (2005). https://doi.org/10.1016/j.renene.2004.06.005
9. Adamo, F., Attivissimo, F., Di Nisio, A., Spadavecchia, M.: Characterization and testing of a tool for photovoltaic panel modeling. IEEE Trans. Instrum. Meas. **60**, 1613–1622 (2011). https://doi.org/10.1109/TIM.2011.2105051
10. Parida, B., Iniyan, S., Goic, R.: A review of solar photovoltaic technologies. Renew. Sustain. Energy Rev. **15**, 1625–1636 (2011). https://doi.org/10.1016/J.RSER.2010.11.032
11. Guechi, A., Chegaar, M.: Effects of diffuse spectral illumination on microcrystalline solar cells. J. Electron Devices **5**, 116–121 (2007)
12. Droz, C., Roux, J., Rouelle, S.B., et al.: Mastering the spectrum in Class A pulsed solar simulators. In: Proceedings of 23rd EUPVSEC, pp. 326–329 (2008)
13. Kenny, R.P., Viganó, D., Salis, E., et al.: Power rating of photovoltaic modules including validation of procedures to implement IEC 61853-1 on solar simulators and under natural sunlight. Prog. Photovoltaics Res. Appl. **21**, 1384–1399 (2013)
14. Sayre, R.M., Cole, C., Billhimer, W., et al.: Spectral comparison of solar simulators and sunlight. Photodermatol. Photoimmunol. Photomed. **7**, 159–165 (1990)
15. Kim, K.A., Dostart, N., Huynh, J., Krein, P.T.: Low-cost solar simulator design for multi-junction solar cells in space applications. In: 2014 Power and Energy Conference at Illinois (PECI). IEEE, pp. 1–6 (2014)
16. Dong, X., Nathan, G.J., Sun, Z., et al.: Concentric multilayer model of the arc in high intensity discharge lamps for solar simulators with experimental validation. Sol. Energy **122**, 293–306 (2015). https://doi.org/10.1016/j.solener.2015.09.004

17. Grandi, G., Ienina, A.: Analysis and realization of a low-cost hybrid LED-halogen solar simulator. In: 2013 International Conference on Renewable Energy Research and Applications (ICRERA). IEEE, pp. 794–799 (2013)
18. Kohraku, S., Kurokawa, K.: A fundamental experiment for discrete-wavelength LED solar simulator. Sol. Energy Mater. Sol. Cells **90**, 3364–3370 (2006)
19. Ferrer, J.P., Martínez, M., Trujillo, P., et al.: Indoor characterization at production scale: 200 kWp of CPV solar simulator measurements. In: AIP Conference Proceedings, AIP, pp. 161–164 (2012)
20. Bickler, D.B.: The calibration of a solar simulator. In: Proceedings of the ASME Solar Energy Applications Committee Winter Annual Meeting (1962)
21. Bisaillon, J.C., Cummings, J.R., Culik, J.S., et al.: Non-traditional light sources for solar cell and module testing. In: Conference Record of the Twenty-Eighth IEEE Photovoltaic Specialists Conference-2000 (Cat. No. 00CH37036). IEEE, pp. 1498–1501 (2000)
22. Namin, A., Jivacate, C., Chenvidhya, D., et al.: Construction of tungsten halogen, pulsed LED, and combined tungsten halogen-LED solar simulators for solar cell-characterization and electrical parameters determination. Int. J. Photoenergy **2012** (2012)
23. Mohan, M.V.A., Pavithran, J., Osten, K.L., et al.: Simulation of spectral match and spatial non-uniformity for LED solar simulator. In: 2014 IEEE Global Humanitarian Technology Conference-South Asia Satellite (GHTC-SAS). IEEE, pp. 111–117 (2014)
24. Sarwar, J., Georgakis, G., LaChance, R., Ozalp, N.: Description and characterization of an adjustable flux solar simulator for solar thermal, thermochemical and photovoltaic applications. Sol. Energy **100**, 179–194 (2014)
25. Dennis, T.: An arbitrarily programmable solar simulator based on a liquid crystal spatial light modulator. In: 2014 IEEE 40th Photovoltaic Specialist Conference (PVSC), pp. 3326–3330 (2014)
26. Dennis, T., Schlager, J.B., Yuan, H.-C., et al.: A novel solar simulator based on a super-continuum laser. In: 2012 38th IEEE Photovoltaic Specialists Conference. IEEE, pp. 1845–1848 (2012)
27. Bari, D., Wrachien, N., Tagliaferro, R., et al.: Reliability study of dye-sensitized solar cells by means of solar simulator and white LED. Microelectron. Reliab. **52**, 2495–2499 (2012)
28. Novickovas, A., Baguckis, A., Mekys, A., Tamosiunas, V.: Compact light-emitting diode-based AAA class solar simulator: design and application peculiarities. IEEE J. Photovoltaics **5**, 1137–1142 (2015). https://doi.org/10.1109/JPHOTOV.2015.2430013
29. Wang, W., Aichmayer, L., Laumert, B., Fransson, T.: Design and validation of a low-cost high-flux solar simulator using fresnel lens concentrators. Energy Procedia **49**, 2221–2230 (2014). https://doi.org/10.1016/j.egypro.2014.03.235
30. Villalva, M.G., Gazoli, J.R., Filho, E.R.: Comprehensive approach to modeling and simulation of photovoltaic arrays. IEEE Trans. Power Electron. **24**, 1198–1208 (2009). https://doi.org/10.1109/TPEL.2009.2013862
31. Meng, H., Xiong, L., He, Y., et al.: Uncertainty analysis of solar simulator's spectral irradiance measurement. In: 6th International Symposium on Advanced Optical Manufacturing and Testing Technologies: Optoelectronic Materials and Devices for Sensing, Imaging, and Solar Energy. International Society for Optics and Photonics, p. 84193A (2012)
32. Meng, Q., Wang, Y., Zhang, L.: Irradiance characteristics and optimization design of a large-scale solar simulator. Sol. Energy **85**, 1758–1767 (2011). https://doi.org/10.1016/j.solener.2011.04.014
33. Yamamoto, M., Ikki, O.: National survey report of PV power applications in Japan. International Energy Agency Co-operative Program on Photovoltaic Power Systems (2010)
34. Dennis, T., Schlager, J.B., Bertness, K.A.: A novel solar simulator based on a supercontinuum laser for solar cell device and materials characterization. IEEE J. Photovoltaics **4**, 1119–1127 (2014)
35. Serreze, H.B., Sobhie, H.M., Hogan, S.J.: Solar simulators-beyond Class A. In: 2009 34th IEEE Photovoltaic Specialists Conference (PVSC). IEEE, pp. 100–105 (2009)

36. Yang, C., Wang, J., Guo, X., et al.: A multisource regular dodecahedron solar simulator structure for uniform flux. IEEE J. Photovoltaics **6**, 516–521 (2016). https://doi.org/10.1109/JPHOTOV. 2015.2504784
37. Chawla, M.K., Tech, P.E.: A Step by Step Guide to Selecting the "Right" Solar Simulator for Your Solar Cell Testing Application. Photo Emiss Tech, Inc, USA (2006)
38. Riedel, N., Pratt, L., Edler, A., Haas, F.: Effects of a neutral density filter in measuring low-light performance with a pulsed light Xe arc solar simulator. In: 2015 IEEE 42nd Photovoltaic Specialist Conference (PVSC). IEEE, pp. 1–4 (2015)
39. Riordan, C., Hulstron, R.: What is an air mass 1.5 spectrum? (Solar cell performance calculations). In: IEEE Conference on Photovoltaic Specialists. IEEE, pp. 1085–1088 (1990)
40. Polly, S.J., Bittner, Z.S., Bennett, M.F., et al.: Development of a multi-source solar simulator for spatial uniformity and close spectral matching to AM0 and AM1. 5. In: 2011 37th IEEE Photovoltaic Specialists Conference. IEEE, pp. 1739–1743 (2011)
41. Shrotriya, V., Li, G., Yao, Y., et al.: Accurate measurement and characterization of organic solar cells. Adv. Funct. Mater. **16**, 2016–2023 (2006)
42. Fanney, A.H., Davis, M.W., Dougherty, B.P., et al.: Comparison of photovoltaic module performance measurements. J. Sol. Energy Eng. **128**, 152–159 (2006)
43. Georgescu, A., Damache, G., Gîrţu, M.A.: Class A small area solar simulator for dye-sensitized solar cell testing. J. Optoelectron. Adv. Mater. **10**, 3003–3007 (2008)
44. Bazzi, A.M., Klein, Z., Sweeney, M., et al.: Solid-state solar simulator. IEEE Trans. Ind. Appl. **48**, 1195–1202 (2012). https://doi.org/10.1109/TIA.2012.2199071
45. Krusi, P., Schmid, R.: The CSI 1000 W lamp as a source for solar radiation simulation. Sol. Energy **30**, 455–462 (1983). https://doi.org/10.1016/0038-092X(83)90116-0
46. Jang, S.H., Shin, M.W.: Fabrication and thermal optimization of LED solar cell simulator. Curr. Appl. Phys. **10**, S537–S539 (2010). https://doi.org/10.1016/j.cap.2010.02.035
47. Bliss, M., Plyta, F., Betts, T.R., Gottschalg, R.: LEDs based characterisation of photovoltaic devices. In: International Conference on Energy Efficient LED Lighting and Solar Photovoltaic Systems Conference, Indian Institute of Technology, Kanpur, India, 27th-29th March (2014)
48. Plyta, F., Mihaylov, B.V., Betts, T.R., Gottschalg, R.: Optical design of a LED solar simulator and survey on its performance characterisation capability. In: Proceedings of the 8th Photovoltaic Science, Application and Technology (PVSAT) January (2012)
49. Plyta, F., Betts, T.R., Gottschalg, R.: Towards a fully LED-based solar simulator-spectral mismatch considerations. In: 28th European Photovoltaic Solar Energy Conference and Exhibition, pp. 3496–3499. (2013) https://doi.org/10.4229/28thEUPVSEC2013-4AV.6.44
50. Georgescu, A., Gîrţu, M.A., Ciupină, V.: Spectral calibration of a LED-based solar simulator—a theoretical approach. J. Optoelectron. Adv. Mater. **15**, 31–36 (2013)
51. Hamadani, B.H., Chua, K., Roller, J., et al.: Towards realization of a large-area light-emitting diode-based solar simulator. Prog. Photovoltaics Res. Appl. **21**, 779–789 (2013). https://doi. org/10.1002/pip.1231
52. Kohraku, S., Kurokawa, K.: New methods for solar cells measurement by LED solar simulator. In: 3rd World Conference on Photovoltaic Energy Conversion, vol. 2, pp. 1977–1980 (2003)
53. Bliss, M., Betts, T.R., Gottschalg, R.: Advantages in using LEDs as the main light source in solar simulators for measuring PV device characteristics. In: Proceedings of SPIE (2008)
54. Tsuno, Y., Kamisako, K., Kurokawa, K.: New generation of PV module rating by LED solar simulator—a novel approach and its capabilities. In: 2008 33rd IEEE Photovoltaic Specialists Conference. IEEE, pp. 1–5 (2008)
55. Krebs, F.C., Sylvester-Hvid, K.O., Jørgensen, M.: A self-calibrating led-based solar test platform. Prog. Photovoltaics Res. Appl. **19**, 97–112 (2011)
56. Kolberg, D., Schubert, F., Lontke, N., et al.: Development of tunable close match LED solar simulator with extended spectral range to UV and IR. Energy Procedia **8**, 100–105 (2011). https://doi.org/10.1016/j.egypro.2011.06.109
57. Kolberg, D., Schubert, F., Klameth, K., Spinner, D.M.: Homogeneity and lifetime performance of a tunable close match LED solar simulator. Energy Procedia **27**, 306–311 (2012). https:// doi.org/10.1016/j.egypro.2012.07.068

58. Plyta, F., Betts, T.R., Gottschalg, R.: Potential for LED solar simulators. In: 2013 IEEE 39th Photovoltaic Specialists Conference (PVSC), pp. 701–705 (2013)
59. Linden, K.J., Neal, W.R., Serreze, H.B.: Adjustable spectrum LED solar simulator. In: Proceedings of SPIE (2014)
60. Luka, T., Eiternick, S., Turek, M.: Rapid testing of external quantum efficiency using LED solar simulators. Energy Procedia **77**, 113–118 (2015). https://doi.org/10.1016/j.egypro.2015.07.018
61. Leary, G., Switzer, G., Kuntz, G, Kaiser, T.: Comparison of xenon lamp-based and led-based solar simulators. In: 2016 IEEE 43rd Photovoltaic Specialists Conference (PVSC), pp. 3062–3067 (2016)
62. Hamadani, B.H., Roller, J., Dougherty, B., Yoon, H.W.: Fast and reliable spectral response measurements of PV cells using light emitting diodes. In: 2013 IEEE 39th Photovoltaic Specialists Conference (PVSC), pp. 73–75 (2013)
63. Dennis, T.: Saturation in solar cells from ultra-fast pulsed-laser illumination. In: 2016 IEEE 43rd Photovoltaic Specialists Conference (PVSC), pp. 3023–3026 (2016)
64. Dudley, J.M., Genty, G., Coen, S.: Supercontinuum generation in photonic crystal fiber. Rev. Mod. Phys. **78**, 1135–1184 (2006). https://doi.org/10.1103/RevModPhys.78.1135
65. Dennis, T., Yasanayake, C., Gerke, T., et al.: A programmable solar simulator for realistic seasonal, diurnal, and air-mass testing of multi-junction concentrator photovoltaics. In: 2016 IEEE 43rd Photovoltaic Specialists Conference (PVSC), pp. 2327–2332 (2016)
66. Mundus, M., Dasa, M.K., Wang, X., et al.: Spectrally shaped supercontinuum for advanced solar cell characterization. In: 31st European Photovoltaic Solar Energy Conference and Exhibition, Hamburg, Germany, pp. 514–519 (2015)
67. Seoul Semiconductor: Product data sheet (2013)
68. Grandi, G., Ienina, A., Bardhi, M.: Effective low-cost hybrid LED-halogen solar simulator. IEEE Trans. Ind. Appl. **50**, 3055–3064 (2014)

Chapter 2
Performance of Solar Cells Using Thermoelectric Module in Hot Sites

M. Benghanem and A. Almohammedi

Abstract Madinah site is considered as one of the most hot cities in the world. In fact, the ambient temperature is between 41 and 55 °C during the summer months. Then, the cell temperature will increase until 83 °C. So, the efficiency of solar cells (SC) will decrease. In this work, we use a thermoelectric module (TEM) for cooling the solar cells in order to get the best performance. The efficiency of solar cells drops by 0.5% per °C rise in temperature. So, we need to keep them at lower temperature to get the best efficiency. The hybrid photovoltaic (PV)–thermoelectric module (PV/TEM) system is suggested for PV applications in hot locations.

Keywords Solar cells performance · Efficiency · Thermoelectric cooler · Hybrid PV/Thermoelectric system

2.1 Introduction

Solar energy is one of the most applicable renewable energy sources in the world. The output power of photovoltaic (PV) system is affected by the incident solar irradiation and the ambient temperature. When temperature of the solar panels increases, its efficiency decreases [1]. To avoid this effect of temperature on solar panels efficiency, it is recommended that the solar panels keep cooled by an appropriate system of cooling.

The thermoelectric (TEC) effect has been used in many different applications such as heating, cooling and generation of electricity. The main application was to generate electricity depending on the difference of temperature between the two terminals of the TEC. Other application is to cool micro-electronic circuits like microprocessors device. For cooling system using TEC, we do not require any fluid or external agent to achieve temperature control. The thermoelectric module (TEM) is a safety component and relatively small in size and weight. The thermoelectric

M. Benghanem (✉) · A. Almohammedi
Physics Department, Faculty of Science, Islamic University, Madinah, Kingdom of Saudi Arabia
e-mail: benghanem_mohamed@yahoo.fr

© The Editor(s) (if applicable) and The Author(s), under exclusive license
to Springer Nature Switzerland AG 2020
A. Mellit and M. Benghanem (eds.), *A Practical Guide for Advanced Methods
in Solar Photovoltaic Systems*, Advanced Structured Materials 128,
https://doi.org/10.1007/978-3-030-43473-1_2

29

module is based on the Seebeck, Peltier and Thomson effects. TEM allow heating, cooling or generating electricity [2]. Cooling integrated circuits in the computer industry have been applied by using TEM to improve computer clock speed below ambient temperatures [3, 4].

Thermoelectric technic has been used as a possible solution as cooling system [5], and the research focuses into cooling microprocessor device with a thermoelectric module using embedded micro-thermoelectric [6]. Recently, some work has been done for solar cooling system. Also, water spray has been used as cooling technic [7, 8]. For cooling the building based on PV system, a TEM device has been realized [9]. To improve the efficiency of PV panels, the temperature has been reached to 10 °C using TEM. Other research work was about using simulation to estimate the behavior of solar cells by incorporating in the software thermoelectric modules. The results of this simulation showed the improvement in the efficiency from approximately 7% up to 11% at 83 °C [10]. Other research has been done using TEM as cooling system to increase the efficiency of PV generator. In fact, the efficiency of solar cells was about 8% before cooling and reached the value of 13 after cooling [11]. The temperature of the solar cells can reach up to 80 °C, which contributes to the deterioration of solar cells and affects all their performance such as life time and efficiency [12]. To minimize the effect of high cell temperature, the researchers tried to cool solar cells in order to obtain the best performance. Other research work proved that the hybrid PV/TEM device gives better results than using the PV only with heat sink module [13].

Saudi Arabia is characterized by high temperature in most cities and approximately all the year except the period from December to February. The ambient temperatures are round 56 °C in summer, and this will give high value of cell temperature and then decrease the efficiency of solar panels. In this present work, we suggest to use the hybrid device PV/TEM, in order to cool the solar panels installed in different applications in Madinah city (KSA).

2.2 Temperature Data Base

In this work, we use the data base collected at Madinah city (Saudi Arabia) with the geographical location (Longitude 39.62° E and Latitude 24.46° N). This location is characterized as semi-arid area with a big potential of solar irradiation which can be reached 8.5 KWh/m^2/day [14].

The database has been collected during the period 2008–2015. The collected data indicates that the ambient temperature vary between 41 and 55 °C during the summer days (Fig. 2.1). The collected data about temperature for the year 2011–2014 is shown in Fig. 2.2. We can see that ambient temperature has been over 50 °C in many days in the period July–August. The rise in ambient temperature allows increasing the cell temperature which affects the performance of solar cells. So, we propose the use of hybrid system PV/TEM which allows fast cooling the solar panel for getting the best performance of solar panel in hot sites such as Madinah city.

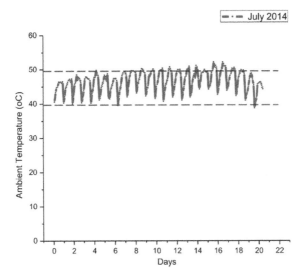

Fig. 2.1 Evolution of temperature at Madinah city (July 2014)

2.3 Thermoelectric Effect

The first application of thermoelectric module (TEM) is to generate electricity due to the difference of temperature, and then, it can transfer heat to electricity.

This effect called Seebeck effect was introduced by Seebeck [15]. The procedure is to apply difference of temperature between two terminals that allow the production of electricity (Fig. 2.3). Also, TEM is considered as a heat pump using Seebeck/Peltier effects. Figure 2.3 shows that two electrical insulator ceramic plates enclose many n and p-type thermal elements that are electrically connected in series and thermally parallel with electrical insulation. TEM is considered as cooling and heating device and could be used for many applications such as generating power and cooling PV module [16, 17]. By using the Peltier effect, TEM can be used for cooling based on the principle of getting heat flux between the junction of two elements P and N [18]. By applying a voltage across two conductors A and B, we will obtain a heat at the junction. The rate dq/dt of the produced heat is

$$\frac{dQ}{dt} = (\pi_A - \pi_B)I \tag{2.1}$$

I is the current, π_A and π_B are Peltier's factors of the conductors.

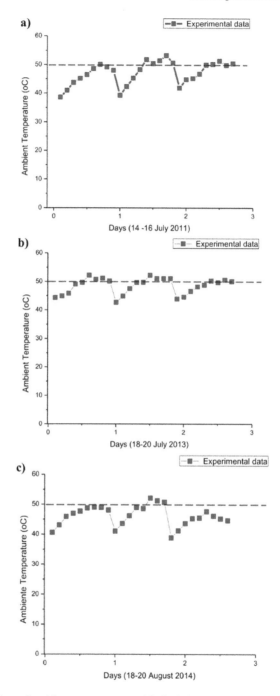

Fig. 2.2 Evolution of ambient temperature at Madinah location. **a** Period: 14–16 July 2011. **b** Period: 18–20 July 2013. **c** Period: 18–20 August 2014

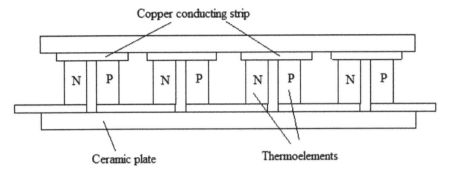

Fig. 2.3 Single stage thermoelectric module

2.4 Modeling of Solar Cell

To study the behaviors of I–V characterization of solar cells, several research works established models to characterize the solar cells [19, 20]. It was demonstrated that the I–V curve of solar cell depends on many parameters such as the series resistor R_S, the shunt resistance R_{SH} and the temperature T. Figure 2.4 illustrates the equivalent circuit of solar cells [21]. From this figure, we deduce that the current I produced by the PV cell is given by the relation:

$$I = I_L - I_D - I_{SH} \tag{2.2}$$

I_L is the photocurrent, I_D is the diode current and I_{SH} is current in the shunt resistance R_{SH}. The voltage V delivered is

$$V = V_j - I \cdot R_S \tag{2.3}$$

V_j is the voltage across shunt resistor, and R_S is the series resistor. I_D is estimated by Shockley diode relation:

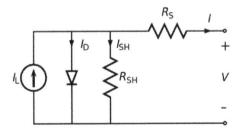

Fig. 2.4 Equivalent circuit of solar cell

$$I_D = I_0 \cdot \left\{ \mathrm{Exp} \left[\frac{q \cdot V_j}{nkT} \right] - 1 \right\} \qquad (2.4)$$

where I_0 is the reverse saturation current, q is the elementary charge, n is the diode ideality factor, k is the Boltzmann's constant, T is the absolute temperature and kT/q is equal to 0.026 V at 25 °C.

The current in the shunt resistance R_{SH} is

$$I_{SH} = \frac{V_j}{R_{SH}} \qquad (2.5)$$

With the Eqs. (2.3) and (2.5), the formula (2.2) becomes

$$I = I_L - I_0 \cdot \left\{ \mathrm{Exp} \left[\frac{q(V + I \cdot R_S)}{nkT} \right] - 1 \right\} - \frac{V + I \cdot R_S}{R_{SH}} \qquad (2.6)$$

The model given in relation (2.6) has given good results in last research [21] to characterize the solar cells.

In this actual research, we have collected the values of R_{SH} and R_S. We have got some results which explain the effect of temperature on R_S and R_{SH}.

2.5 Hybrid System Solar Cell/Thermoelectric Module (SC/TEM)

Figure 2.5 shows the experimental device SC/TEM realized as follows:

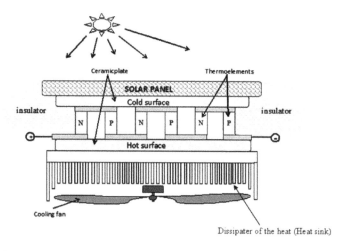

Fig. 2.5 SC/TEM cooling system

The cold surface of the TEM is attached to the backside of the solar cell, while the hot surface is attached to a heat sink to enhance heat extraction. The cell temperatures reached the values in the range of 60–80 °C due to the exposer of the solar cells to solar radiation for several hours. So, this hybrid system allows cooling the solar cells.

2.6 Results

The impact of temperature on the behavior of solar cells has been discussed in this present experiment. We have proposed and used the hybrid SC/TEM system for cooling the solar cells as shown in Fig. 2.6.

2.6.1 Experimental Measurements

Experimental measurement is based first on the realization of hybrid SC/TEM system, and second step is about measurement of $I–V$ characteristic before and after cooling. The used instrument is as follows:

- Source meter for measuring the characteristic of solar cell by exposing it at a standard solar. We have got the $I–V$ curve with the values of some parameters related to the tested solar cell such as serial resistance, fill factor and its efficiency.

Fig. 2.6 Experimental setup of SC/TEM cooling system

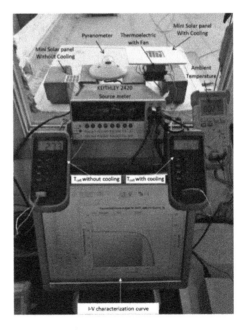

- Tracer software for measuring and analyzing the data from I–V curves using some calculation technics.
- Pyranometer for solar irradiation measurement.
- Thermocouple (type K) to collect measurement about the ambient temperature and cell temperature.

First step in this experiment consists of measuring all parameters of two identical PV panels (Fig. 2.7a, b).

The second part consisted to measure the ambient and cell temperature before and after cooling. We have repeated measurement many times for the same panels in order to ensure the obtained measure. Figure 2.8 shows the effect of cooling on the value of cell temperature. This results allow improving the performance of the used PV panels.

Fig. 2.7 **a** I–V characteristic for mini-panel 1. **b** I–V characteristic for mini-panel 2

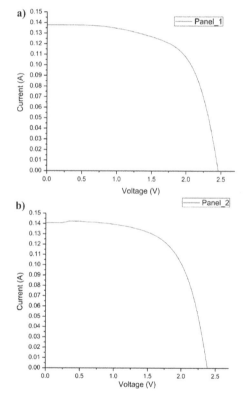

Fig. 2.8 Ambient and cell temperature before and after cooling

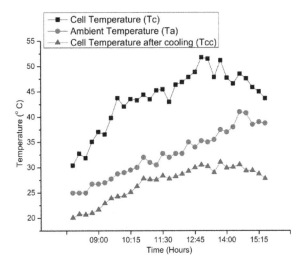

2.6.2   *Effect of the Temperature on* **I–V** *Characteristic*

The *I–V* characteristic is affected by the temperature. If *T* increases, the magnitude of the exponent (Relation 2.6) is reduced, and the saturation current I_0 increases with *T*. The produced photocurrent I_L increases with increasing temperature. Also, from the above parameters, we can estimate the impact of temperature on solar cell efficiency. In fact, the variation in voltage is more apparent than the variation in current due to the temperature. Then, the effect on efficiency is the same to that voltage. Figure 2.9 illustrates the impact of temperature on *I–V* curves.

Fig. 2.9 Impact of temperature on *I–V* characteristic

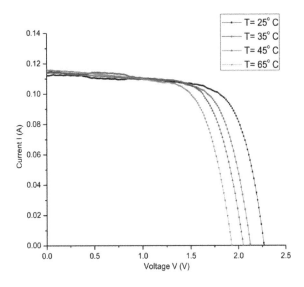

2.6.3 Impact of Temperature on Series and Shunt Resistances

The series resistor is an important factor which should be smaller for the best solar cells. In this experiment, we have studied the impact of temperature on the value of serial resistor. We have observed from measurement that when the temperature increases, the series resistance increases also. This will decrease the output voltage as mentioned in the relation (2.3). The parallel resistor is an important factor which should be greater for the best solar cells. From our experimental measurement, we observed that when the temperature increases, the shunt resistance decreases. This allows reduction in open circuit voltage as shown in Fig. 2.8. The measured data of series resistances has been plotted in Fig. 2.9 before and after cooling. We can observe from this figure that we have got good results since the values of R_S are smaller after cooling. Also, Fig. 2.10 shows that R_S decreases if the cell temperature decreases. Also, Fig. 2.11 shows that the measure of shunt resistances after cooling is higher. So, cooling the solar cell allows to get a decrease in series resistance and an increase in shunt. By the way, we have interested to check what about efficiency of the solar cell after cooling. In fact, Fig. 2.12 shows that the measured efficiency increases after cooling. Figure 2.13 illustrates the increase in efficiency % per °C decrease in temperature. The efficiency has been estimated and represented in Fig. 2.14. The increase in efficiency (%) per °C decrease in temperature is shown in Fig. 2.15 which show that the maximum increase is 1.3%. This is a good result as we can also show in Fig. 2.16 which represent the evolution of efficiency vs. cell temperature.

Power produced with combined SC/TEM system is plotted in Fig. 2.17. The pick power is 182 mW at 942 W/m² at 25 °C. This explains that the power produced by SC using cooling system is higher than the power produced without cooling system (Fig. 2.18).

Fig. 2.10 Effect of series and shunt resistance on the *I–V* characteristic

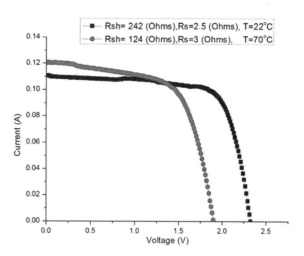

Fig. 2.11 Series resistance before and after cooling

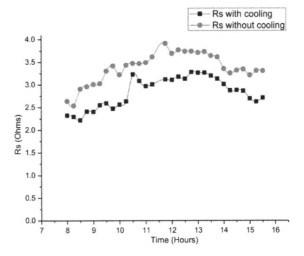

Fig. 2.12 Evolution of series resistance with cell temperature

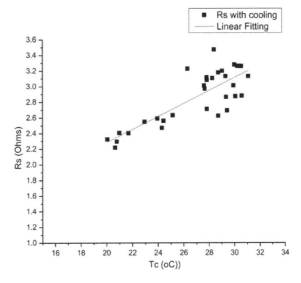

2.7   Hybrid PV Panel/TEM System Proposed

Figure 2.18 illustrates the new proposed cooling system for PV panels using TEM. This system could be applied for solar panels installed in hot locations such as Madinah city. Also, for any application of PV system in hot sites, we propose to use the new hybrid PV/TEM system (Fig. 2.19) for cooling the PV systems and getting the best performance. Also, it is more suitable to add an extra panel just for powering the TEM used in the PV system. In fact, the output voltage of the extra panel increases when the solar irradiation increases and then the cold side of each

Fig. 2.13 Evolution of shunt resistance before and after cooling

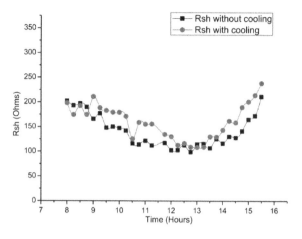

Fig. 2.14 Efficiency before and after cooling

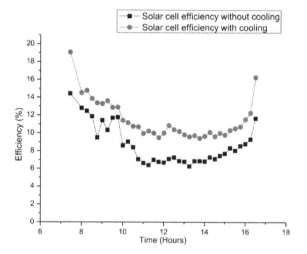

TEM became more colder by the increase of output voltage as indicated in Fig. 2.20. This allows decreasing the cell temperature of panels used in PV system which will get the best performance.

2.7.1 Economic Analysis

The aim of this analysis is to find the total cost of the proposed hybrid PV/TEM system by comparison to traditional PV system without cooling system. The economic analysis takes into account a number of parameters like the location and wattage cost. The mechanism to get the economic analysis is [22]:

Fig. 2.15 Evolution of
efficiency

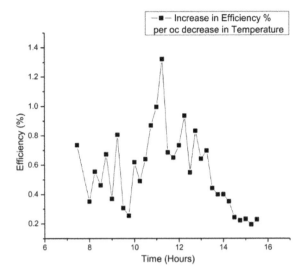

Fig. 2.16 Efficiency of solar
cell versus cell temperature

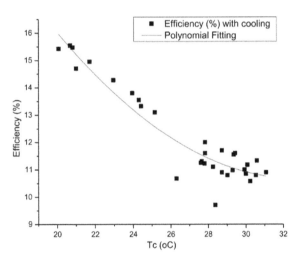

- Energy demand calculation (Q).
- Number of sunlight hours in the given location (H).
- Calculation of the lowest solar irradiation in this location (I).
- Determination of the capacity of PV system (PV_C):

$$PV_C = \frac{Q \cdot PE_P}{H \cdot E_P} \tag{2.7}$$

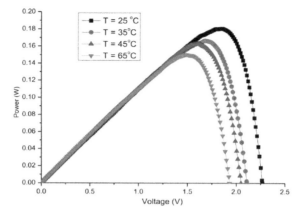

Fig. 2.17 Power generated with solar cell/thermoelectric module (SC/TEM) system

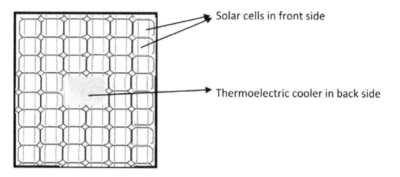

Fig. 2.18 PV panel/thermoelectric module (TEM) cooler system

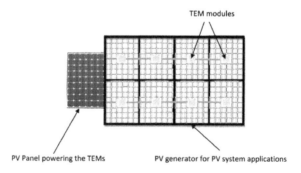

Fig. 2.19 Hybrid PV/TEM system

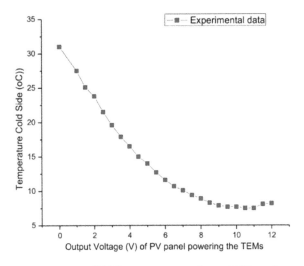

Fig. 2.20 Cold side temperature of TEMs versus output voltage of PV panel

E_P represents the efficiency of the PV system, and PE_P is the demand of energy needs.

The area of PV system (PV_A) is

$$PV_A = \frac{PV_C \cdot H}{I} \tag{2.8}$$

The cost of PV system (C_{PV}) is

$$C_{PV} = PV_C \cdot 1000 \cdot C_W \tag{2.9}$$

where C_W is the wattage cost.

The total cost of hybrid PV/TEM system is

$$C_{Tot} = C_{PV} + C_{TEM} \tag{2.10}$$

where C_{TEM} is the extra price of TEM requested. The price of electricity generated by a PV system depends on the initial price [23]. The PV system price depends on the local market. The price of photovoltaic system is in the range 1–2 \$/W for small size. The extra price of hybrid PV/TEM system is in the range of 3.1–11.8%. So, the average of the extra price is approximately 6% for small-scale PV. For large scale, the extra price will be smaller as indicated in Fig. 2.21.

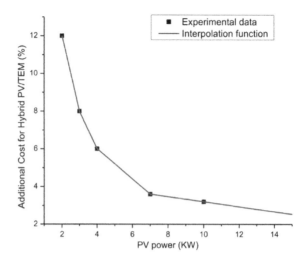

Fig. 2.21 Additional cost of proposed hybrid PV/TEM system versus PV power

2.7.2 Example

The price of 4 KW/24 V off grid PV system is 5800 $ [24]. Twelve solar panels of 300 W are necessary. So, we request 12 TEMs which cost 12×13 US$ = 156 US$.

The price of the PV panel which powering the 12 TEMs is 200 US$. So, the global price of extra accessories needed for hybrid PV/TEM system is 356 US$. Thus, this means that the price of TEM and the solar panel powering the TEMs is about 6% of the global price.

2.8 Conclusion

In Madinah location, the cell temperature has reached the value of 83 °C. The results indicate that the hybrid system SC/TEM can be operated at 64 °C of cell temperature without loss.

The suggested hybrid PV/TEM system allows increasing the efficiency of solar cells. The TEM is used as cooling system, and it is attached on the backside of the solar cell. The performance of solar cells has been improved by increasing their efficiency by 0.5% per °C decrease in temperature.

The PV/TEM system suggested for photovoltaic applications in hot locations gives the best performances, while the extra price is about 6% of the global price of traditional PV systems.

Acknowledgements We would like to thank the deanship of scientific research at Islamic University (Madinah, KSA) for supporting this first (Tamayouz) program of academic year 2018/2019, research project No.: 1/40. All collaboration works are gratefully acknowledged.

References

1. Tang, X., Quan, Z., Zhao, Y.: Experimental investigation of solar panel cooling by a novel micro heat pipe array. Energy Power Eng. **2**(3), 171–174 (2010)
2. Gould, C.A., Shammas, N.A.Y., Grainger, S., Taylor, I.A.: Comprehensive review of thermoelectric technology, micro-electrical and power generation properties. In: Proceedings of 26th IEEE International Microelectronics Conference, Nis, Serbia, 2 May 2008, pp. 329–332
3. Sharp, J., Bierschenk, J., Lyon, H.B. Jr.: Overview of solid-state thermoelectric refrigerators and possible applications to on-chip thermal management. Proc. IEEE **9**(8), 1602e1612 (2006)
4. Future Publishing Ltd: The Ultimate Overclocking Handbook, pp. 11–19. Future Publishing Ltd, UK (2009)
5. Mahajan, R., Chiu, C., Chrysler, G.: Cooling a microprocessor chip. Proc. IEEE **94**(8) (2006)
6. Snyder, G.J., Soto, M., Alley, R., Koester, D., Conner, B.: Hot spot cooling using embedded thermoelectric coolers. In: Proceedings of 22nd IEEE Semiconductor Thermal Measurement and Management Symposium, March 2006, pp. 135–143
7. Moharram, K.A., Abd-Elhady, M.S., Kandil, H.A., El-Sherif, H.: Influence of cleaning using water and surfactants on the performance of photovoltaic panels. Energy Convers. Manag. **68**, 266–272 (2013)
8. Mavroidis, C., Hastie, J., Grandy, A., Anderson, M., Sweezy, A., Markpolous, Y.: Robotic Device for Cleaning Photovoltaic Panel Arrays. Department of Mechanical and Industrial Engineering, Northeastern University, Green Project-Sustainable Technology and Energy Solutions. Patent Number 61/120097 (2009)
9. Kane, A., Verma, V.: Performance enhancement of building integrated photo-voltaic module using thermoelectric cooling. Int. J. Renew. Energy Res. **3**(2), 320–324 (2013)
10. Ahadi, S., Hosein, H.R., Faez, R.: Using of thermoelectric devices in photovoltaic cells in order to increase efficiency. Indian J. Sci. Res. **2**(1), 20–26 (2014)
11. Borkar, D.S., Prayagi, S.V., Gotmare, J.: Performance evaluation of photovoltaic solar panel using thermoelectric cooling. Int. J. Eng. Res. **3**(9), 536–539 (2014)
12. Yang, D., Yin, H.: Energy conversion efficiency of a novel hybrid solar system for photovoltaic, thermoelectric, and heat utilization. IEEE Trans. Energy Convers. **26**(2), 662–670 (2011)
13. Simons, R.E., Chu, R.C.: Application of Thermoelectric Cooling to Electronic Equipment: A Review and Analysis, 3rd edn., vol. 2, pp. 68–73. Clarendon, Oxford (1892)
14. Benghanem, M.: Measurement of meteorological data based on wireless data acquisition system monitoring. Appl. Energy **86**(12), 2651–2660 (2009)
15. Yamashita, O.: Resultant Seebeck coefficient formulated by combining the Thomson effect with the intrinsic Seebeck coefficient of a thermoelectric element. Energy Convers. Manag. **50**, 2394–2399 (2009)
16. Choi, J.S., Ko, J.S., Chung, D.H.: Development of a thermoelectric cooling system for a high efficiency BIPV module. J. Power Electron. **2**, 187–193 (2010)
17. Wu, C.: Analysis of waste-heat thermoelectric power generators. Appl. Therm. Eng. **16**(1), 63–69 (1996)
18. Min, G., Rowe, D.M.: Improved model for calculating the coefficient of performance of a Peltier module. Energy Convers. Manag. **41**, 163–171 (2000)
19. Benghanem, M., Alamri, S.: Modeling of photovoltaic module and experimental determination of serial resistance. J. Taibah Univ. Sci. **2**, 94–105 (2009)

20. Fornies, E., Balenzategui, J.L., Alonso-García, M.D.C., Silva, J.P.: Method for module R_{sh} determination and its comparison with standard methods. Sol. Energy **109**, 189–199 (2014)
21. Benghanem, M., Alamri, S., Mellit, A.: New IV characterization model for photovoltaic modules and experimental determination of internal resistances. Mediterr. J. Meas. Control **5**(1), 22–30 (2009)
22. Tzouanas, C.N., Tzouanas, V.: Study of a photo-voltaic (PV) system using excel: economic analysis, modeling, simulation, and optimization. Paper presented at 2012 ASEE Annual Conference, San Antonio, Texas, June 2012. https://peer.asee.org/21958
23. Amber, M., Pickrell, K., DeBenedictis, A., Price, S.: Cost Effectiveness of Rooftop Photovoltaic Systems for Consideration in California's Building Energy Efficiency Standards. California Energy Commission, Energy and Environmental Economics, Inc. Publication Number: CEC-400-2013-005-D (2011)
24. http://sunelec.com/pv-systems/off-grid-system

Chapter 3
Optical Optimization of Tandem Structure Formed by Organic Photovoltaic Cells Based on P3HT: PCBM and PBBTDPP2: PCBM Interpenetrating Blends

Z. Abada and A. Mellit

Abstract The chapter focuses on the optical optimization of the tandem structure composed of organic photovoltaic (OPV) cells based on interpenetrating blend materials P3HT: PCBM and pBBTDPP2: PCBM. For this purpose, a simulation based on the transfer matrix formalism is developed. The aim is to calculate the current (J_{SC}) (assuming 100% internal Quantum efficiency). Optical performance of the two OPV cells separately is studied. Firstly, the P3HT: PCBM OPV cell is optimized to find the geometry of the stack giving the best optical properties. The OPV cell structure is given by: glass/ITO/PEDOT: PSS/P3HT: PCBM/Ca/Al. The best J_{SC} current value (in this case 12.48 mA/cm^2) is reached for an active layer thickness $d_{\text{Active layer}} = 91$ nm. In a second time, the optical yield of the OPV cell based on pBBTDPP2: PCBM interpenetrating blend is optimized. A comparison of the optical performance of this cell with those of OPV cells based on materials commonly used in the field of OPVs is carry out. The simulation shows a pBBTDPP2: PCBM interpenetrating blend OPV cell offering the best result ($J_{SC} = 17$ mA/cm^2) in addition of a high absorption in the 600–800 nm range. Finally, the tandem structure formed by the two OPV cells is studied taking into account the arrangement of the two cells with respect to each other. Two configurations are considered namely, the Normal Tandem Solar Cell (NTSC) which considers P3HT: PCBM blend layer as the front active layer and pBBTDPP2: PCBM blend layer as the back active layer and the Reverse Tandem Solar Cell (RTSC) which considers pBBTDPP2: PCBM blend layer on top of the tandem structure and P3HT: PCBM blend layer on the back of the device. The aim is to find for both configurations the optimized thicknesses of the blend active layers of the front and rear OPV cells giving the highest current matching. Results show that NTSC configuration is more efficient for the large thicknesses of the top cell, whereas the RTSC configuration is more efficient for the thin thicknesses of the top cell.

Z. Abada (✉) · A. Mellit
Renewable Energy Laboratory, Jijel University, Jijel, Algeria
e-mail: zakelec2015@gmail.com

© The Editor(s) (if applicable) and The Author(s), under exclusive license
to Springer Nature Switzerland AG 2020
A. Mellit and M. Benghanem (eds.), *A Practical Guide for Advanced Methods in Solar Photovoltaic Systems*, Advanced Structured Materials 128,
https://doi.org/10.1007/978-3-030-43473-1_3

Keywords Optical modeling · Organic solar cell · P3HT: PCBM blend ·
pBBTDPP2: PCBM blend · Tandem structure · Current matching

3.1 Introduction

Organic solar cells or OPVs (Organic Photovoltaics) based on the concept of
Polymer-Fullerene bulk heterojunction (BHJs) are particularly interesting because
of the multiple advantages offered in terms of low cost, lightness and flexibility
combined with a simple and economical manufacturing process. This structure over-
comes the limitation of active layer thickness caused by the low value of the exciton
diffusion in bilayer OPV cells [1–3].

However, and for commercial reasons, organic photovoltaic solar cells need to
reach efficiencies above 10% and achieve lifetimes of several thousands of hours.
These two parameters (lifetime and efficiency) constitute the main constraint on the
large-scale deployment of organic panels.

To overcome the problem of the low OPV cell efficiencies, mainly caused by the
low charge mobility in organic materials, tandem OPV cells are used. These latter are
promising candidates to reach both high efficiencies and long lifetime. The tandem
OPV cell is assembled by two OPV cells, one above the other. Each one absorbs in
a different light wavelength range so as to expand the absorption spectrum of the
structure thus improving efficiency with respect to each independent cell [4–7].

The design of this kind of cells being delicate, it is necessary to play on many
parameters: optical and electrical parameters of the materials forming the active
layers of the tandem OPV cells without forgetting the process of implementation.
However, given the huge range of new organic materials with high optical efficiency
that can be combined to arrive at high efficiency tandem structures, optical modeling plays a very important role in the design of such structures before moving to
technological achievement. Indeed, it allows us to have an overview of the optical
properties of the tandem structure. These properties are in most cases out of reach
of direct experimental measurements such as the profile of the electromagnetic field
$|E|^2$ which provides valuable and precise indications for the technological realization
of these cells.

Given the very thin aspect of the layers forming the stack of OPV cells constituting
the structure of the tandem (a few hundred nanometers), optical interference are
generated in the stack. Thereby, the electromagnetic field distribution inside the OPV
device strongly depends on layer thicknesses. So, the one-dimensional transfer matrix
formalism [8–12] is applied to model the distribution of this field inside the OPV
cell. The aim is to adjust the stack geometry so that the field is maximum in the active
layer. From the distribution of the electromagnetic field, the rate of exciton generation
is computed. The latter is used to calculate the short-circuit current generated in the
OPV cell.

The interest of this chapter is focused on the optical optimization of the P3HT: PCBM/pBBTDPP2: PCBM tandem OPV cell. To do this, the following structure has been adopted:

- Firstly, the optimized P3HT: PCBM interpenetrating blend OPV cell (using optical modeling) is presented.
- Then, a comparison of the optical performance of the pBBTDPP2: PCBM blend OPV cell with cells based on materials commonly used in the field of organic photovoltaics is executed to show the potential offered by the pBBTDPP2: PCBM blend OPV cell and its ability to be used in a tandem structure with the P3HT: PCBM blend OPV cell. The optimized pBBTDPP2: PCBM blend OPV cell is then defined.
- Finally, the P3HT: PCBM/pBBTDPP2: PCBM tandem OPV cell is optimized taking in account the arrangement of the two cells with respect to each other (NTSC and RTSC configurations). Seen that the two subcells are mounted in series, the simulation aims to find the best configuration giving the highest current matching (the most important criterion in such structures).

3.2 Optical Optimization of OPV Cells Based on P3HT: PCBM Interpenetrating Blend

3.2.1 P3HT: PCBM OPV Cell Presentation

In this section, the interest is focused on thicknesses optimization of the OPV cell based on P3HT: PCBM blend. The device structure is a thin-film stack (layers thickness is often of the order of one hundred nanometers) which consists of a glass/ITO/PEDOT: PSS/P3HT: PCBM/Ca/Al structure. (P3HT: PCBM refers to poly(3-hexylthiophene): 6,6-phenyl C61-butyric acid methyl ester and PEDOT: PSS refers to poly (3,4-ethylene dioxythiophene) doped with poly (styrene sulfonate)). Where:

- **ITO** is a transparent electrode layer placed before the organic one.
- **PEDOT: PSS** layer improves the surface rugosity and charge injection and removes the direct contact between the oxide and the organic photoactive layer which can be harmful.
- **P3HT: PCBM** interpenetrating blend layer is the organic photoactive layer. It's a mixture by volume of two semiconductors more or less imbricate one in the other. P3HT material plays the role of donor semiconductor and PCBM is the acceptor one. The specificity of this structure is the multiplication by volume of donor/acceptor interfaces in such a way that all excitons will be diffused without recombining to a donor/acceptor interface and dissociates whatever the position of creation of the exciton.

- **Ca** layer is a thin interfacial layer. It is deposited before the cathode layer. Its role is to reduce the exciton dissociation at the interface between the organic semiconductor and the metal layer.
- **Al** layer is a Thick metal electrode at the end of the stack used to reflect the light rays transmitted in the stack and to cross again the active layer. Without this metal layer absorption would be reduced by more than 30%.

3.2.2 Simulation Software Presentation

Inspired from McGehee Group Matlab software [13], the Matlab program developed for this purpose uses the transfer matrix formalism with two subsets of 2×2 matrices to calculate optical interference and absorption in multilayer stacks [8, 9]. The program requires knowledge of the wavelength dependent complex index of refraction of each material in order to calculate the interference of coherent reflected and transmitted waves at each interface in the stack. It should be noted that the imaginary part of the complex index of refraction, k, is related to the extinction coefficient and is responsible for absorption in a medium. The real part, n, determines the wavelength of light of a given energy in a material and is important for calculating where areas of constructive and destructive interference occur [14].

The software needs to know names of materials making up the structure of the cell and their respective thicknesses (in nm). The optical constants data (n and k) of these materials are listed in a file called Index_of_Refraction_library.xls [13]. The software calculates optical properties of the stack, namely transfer matrices, normalized electric field intensity $|E|^2$, dissipated energy profile (Q), generation rate (G) and J_{SC} under AM 1.5 illumination (assuming 100% internal quantum efficiency) at all wavelengths.

It should be noted that the software takes into account:

- A normal light incidence.
- A neglected optical interference in glass substrate. (The glass substrate is thicker than the coherence length of light).
- Internal quantum efficiency IQE equal to unity (100%).

3.2.3 Simulation

The aim of the simulation is to optimize the geometry of the OPV cell leading to the best optical efficiency. Thickness optimization consists of placing the maximum of the distribution of the electromagnetic field in the vicinity of the photoactive layer [8]. This method increases optical absorption (η_A) in the photoactive layer. So, when supposing that internal quantum efficiency is equal to unity, maximizing the distribution of the electromagnetic field in the photoactive layer amounts to maximize the short-circuit current passing through this layer.

As mentioned above, the OPV cell studied is given by the following structure: Glass/ITO/PEDOT: PSS/P3HT: PCBM/Ca/Al. The thickness values of ITO and Al layers are fixed to those used experimentally, i.e., 180 nm for ITO layer [10, 15] and 100 nm for Al layer [10, 16], and the thicknesses of the PEDOT: PSS and P3HT: PCBM layers are varied: PEDOT: PSS layer between 20nm and 200 nm and P3HT: PCBM layer between 20nm and 120 nm. The minimum value is set to 20 nm to ensure homogeneity of the active layer, and the maximum values set to 120 nm to have a layer thickness less than the charge conduction length L_{CC} in P3HT: PCBM blend layer. The simulation proposes to vary these thicknesses simultaneously and to collect the optimal thicknesses giving the best values of J_{SC} and Q. The optimized thicknesses are given by: $d_{PEDOT: PSS} = 27$ nm and $d_{Active\ layer} = 91$ nm giving the best results: $Q = 304$ W m^{-2} and $J_{SC} = 12.48$ mA/cm^2.

The optical properties of the optimized OPV cell are then plotted in Fig. 3.1.

This figure shows that the electric field is maximized and presents a good distribution over the entire thickness of the active layer Fig. 3.1a, b. It shows also a good absorption in this region ($|E|^2 < 1$). Figure 3.1c shows also a good distribution of the exciton generation rate over the entire thickness of the active layer. The results are close to those available in the literature [8, 11].

For more details, the optical optimization of the cell presented above as well as the simulation method adopted (transfer matrix formalism) is presented in our work [17].

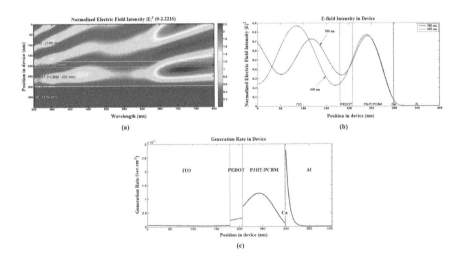

Fig. 3.1 Optical properties of P3HT: PCBM OPV cell. **a** Normalized electric field $|E|^2$ distribution as a function of depth and wavelength. **b** E field intensity for discrete wavelengths. **c** Exciton generation rate in device

3.3   PBBTDPP2: PCBM Blend OPV Cell Optical Properties

3.3.1   *Material Presentation*

poly[3,6-bis-(40-dodecyl-[2,   20]   bithiophenyl-5-yl)-2,5-bis-(2-ethyl-hexyl)-2,5-dihydropyrrolo[3,4-]pyrrole-1,4-dione] (*pBBTDPP2*) *is a* very promising organic material. It shows a low optical bandgap of about 1.4 eV in thin films. The absorption range of pBBTDPP2: PCBM blend layer is extended to reach 860 nm with o-dichlorobenzene as solvent [7]. Thus, a tandem solar cell based on P3HT: PCBM and pBBTDPP2: PCBM almost covers the whole UV and visible parts of the solar spectrum.

In order to show the potential offered by pBBTDPP2: PCBM blend OPV cell, its optical performances were compared to those of cells based on materials commonly used in the field of organic photovoltaics. The goal being to show the complementarity between P3HT: PCBM and pBBTDPP2: PCBM blend OPV cells to widen the absorbed light range to the entire visible spectrum and thus to improve the optical efficiency of OPV cell structure.

All materials chosen for the comparison study have as acceptor material the [6]-phenyl C61 butyric acid methyl ester) **PCBM** and as donor material the following materials:

- poly-[2-(3,7-dimethyloctyloxy)-5-methyloxy]-para-phenylene-vinylene] **(MDMO-PPV)**.
- poly(9,9 0-dioctyl-2,7-fluorine diylvinylene-co-2,5-thiophene) **(PFTBT)**.
- poly[N-9′-heptadecanyl-2,7-carbazole-alt-5,5-(4′,7′-di-2-thienyl-2′,1′,3′-benzothiadiazole)] **(PCDTBT)**.
- poly[3,6-bis(40-dodecyl-[2,   20]bithiophenyl-5-yl)-2,5-bis(2-ethyl-hexyl)-2,5-dihydropyrrolo[3,4-]pyrrole-1,4-dione] **(pBBTDPP2)**.
- poly(3-hexylthiophene) **(P3HT)**.

These blend materials namely MDMO-PPV: PCBM (1:4), PFTBT: PCBM, PCDTBT: PCBM and pBBTDPP2: PCBM materials were added to the database of the simulation software. (P3HT: PCBM material already exists). Data of these materials, i.e., n and k: real and imaginary parts of the complex index of refraction have been derived from the literature [6, 18–20].

It should be noted that the optical constant characteristics of the active layers cited above can be divided into two parts. The range between 300 nm and 350 nm: its tendency follows that of the PCBM material and the range between 350 nm and 800 nm: its tendency follows that of the donor material [21]. It is also noted that pBBTDPP2: PCBM (1:2) is the only material among the five chosen to offer a high absorption in the 600–800 nm range. Other active layers are not being able to absorb at wavelengths greater than 650 nm.

Table 3.1 Optimized OPV cell for different donor materials of the blend active layer

	PEDOT : PSS optimized thickness (nm)	Active layer optimized thickness (nm)	J_{sc} (assuming 100% internal quantum efficiency) (mA/cm^2)
MDMO-PPV: PCBM (1:4)	20	93	07.42
PFTBT: PCBM	20	92	08.22
P3HT: PCBM	27	91	12.29
PCDTBT: PCBM	21	90	12.41
pBBTDPP2: PCBM	**20**	**120**	**17**

The data in bold correspond to the optimized thicknesses of the OPV cell (in this case pBBTDPP2: PCBM blend OPV cell) giving the best short-circuit current Jsc

3.3.2 Simulation

Returning to our simulation, the optical optimization of the ITO/PEDOT: PSS/Active layer/Ca/Al structure is firstly carried out for each of the blend materials cited above by following the same procedure used in the first section to optimize thicknesses of the OPV cell based on P3HT: PCBM interpenetrating blend.

Indeed, ITO, Ca and Al layers having fixed values, the PEDOT layer and the active layer thicknesses are varied in order to recover those giving the higher short-circuit current value. The PEDOT: PSS layer is varied between 20 nm and 70 nm, and the active layer between 20 nm and 120 nm to avoid problems related to the low mobility of carriers in the organic thin film. The thickness values of ITO, Ca and Al layers are fixed to those commonly used experimentally (i.e., 180 nm for ITO layer, 1 nm for Ca layer and 100 nm for Al layer). Simulations were carried out for a wavelength range of between 350 nm and 800 nm. Table 3.1 shows the optimized OPV cells having as active layer the five materials mentioned above.

The results obtained are then used to calculate the fraction of light absorbed in the active layer for each of the optimized cells. The results of the simulation are shown in Fig. 3.2. These results strongly support those obtained in Table 3.1.

3.3.3 Discussion

It should be noted that the absorption in the active layer at any wavelength depends mainly on the corresponding extinction coefficient and refractive index. Indeed, the higher the $\Delta n = n_{Active\ layer} - n_{PEDOT}$ is, the less light is transmitted to the active layer. Therefore, the higher the extinction coefficient, the more light is absorbed. By comparing the results obtained for a single wavelength (e.g., $\lambda = 480$ nm), it should be noted that the best absorption rate is obtained for the P3HT: PCBM active layer having as optical constants $n = 1.73$ ($\Delta n = 1.73-1.52 = 0.21$) and $k = 0.326$ (for

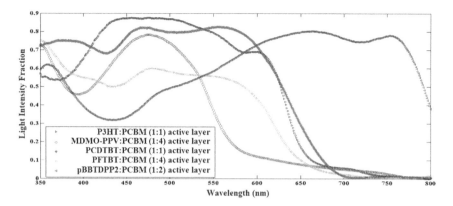

Fig. 3.2 Fraction of light absorbed in the active layer for different donor materials of the blend active layer used in simulation

this wavelength). For the same wavelength, the lowest absorption rate is given by the pBBTDPP2: PCBM active layer having as optical constants $n = 1.749$ ($\Delta n = 1.749$–$1.52 = 0.229$) and $k = 0.0738$. It should be noted that for this wavelength, the two layers have comparable refractive indexes. However, the difference between their extinction coefficients is significant.

Similarly, the overall results (over the entire visible spectrum) can be explained. Table 3.2 presents the mean values of the optical constants of the active layers used in this study (over a wavelength range between 350 nm and 800 nm). Thus, the low optical efficiency of the MDMO-PPV: PCBM active layer (Fig. 3.2 and Table 3.2) can be explained by the combined effect of a low mean value of the extinction coefficient ($k = 0.100021$ is the lowest mean value relative to those of other layers) and a relatively high Δn. This leads to a weak light transmission followed by a low absorption of the latter (Table 3.2).

The PCDTBT: PCBM active layer offers the best average value of the extinction coefficient. However, its relatively high refractive index decreases its optical efficiency with respect to the pBBTDPP2: PCBM active layer which has, in addition to

Table 3.2 Average values (over the entire visible spectrum) of optical constants for active layers used in simulation

	n (average value)	$\Delta n = n_{\text{Active layer}} - n_{\text{PEDOT}}$ (average value)	K (average value)
MDMO-PPV: PCBM	1.937997	0.457357	0.100021
PFTBT: PCBM	1.987333	0.506693	0.116304
P3HT: PCBM	1.896896	0.416256	0.184112
PCDTBT: PCBM	1.961119	0.480479	0.21629
pBBTDPP2: PCBM	1.815154	0.334514	0.206749
PEDOT: PSS	1.48064	0	0.043554

a relatively high mean value of the extinction coefficient, the lowest mean value of Δn, thus promoting the best transmission of light to the active layer (Table 3.2). This clearly explains the results found in Table 3.1 (the best result is given by the pBBT-DPP2: PCBM active layer). In addition, pBBTDPP2: PCBM active layer offers the feature of absorbing light over the entire visible spectrum with a high absorption in the 600–800 nm range (Fig. 3.2). Other active layers are not being able to absorb at wavelengths greater than 650 nm (This is the case for most organic semiconductor materials).

In conclusion, it can be said that the OPV cell based on pBBTDPP2: PCBM blend material is well placed to be used in a tandem structure with the OPV cell based on P3HT: PCBM blend material. Indeed, this will extand the range of absorbed light to the entire visible spectrum and improve the optical efficiency of the structure.

3.4 Tandem Structure Based on P3HT: PCBM and PBBTDPP2: PCMB Blend Materials

Tandem OPV cell structure is a good solution to overcome the limitations caused by the narrow absorption range in OPV solar cells. In such structures, it is necessary to ensure the current matching of both subcells of the tandem structure [7]. Indeed, to achieve high efficiency tandem polymer OPV cells, it is essential to have a high performance low bandgap polymer for the rear cell to match the current of the front cell. Because the two sub-cells are connected in series, the overall current of the tandem device will be limited by the sub-cell with lower current. Thus, the most important criterion for the rear OPV cell is current matching, which means that it can provide a same photocurrent as the front cell when applied into the tandem structure [22].

3.4.1 Tandem Structure Configurations

As seen in the previous section, the complementarity between the two blend OPV cells studied pBBTDPP2: PCBM and P3HT: PCBM in the absorption of the light spectrum makes them very suitable for tandem cell application. So, in this section, the interest is focused on the study of the optical optimization of this kind of structures.

Many publications have treated this structure [6, 7] in order to find the optimized geometry leading to the best optical properties of the device. Two models of tandem structure are studied (see Fig. 3.3):

The conventional tandem solar cell in which materials mainly absorbing light of shorter wavelengths acts as the front active layers to provide a window for the back cell, while materials mostly absorbing light of longer wavelengths work as the

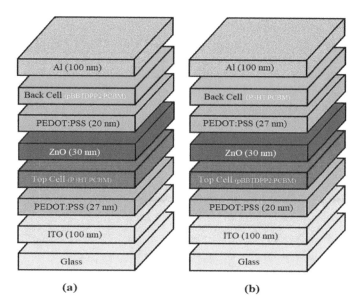

Fig. 3.3 Schematic diagram of the simulated OPV tandem cell. **a** Normal Tandem Solar Cell (NTSC). **b** Reverse Tandem Solar Cell (RTSC)

back active layers. This structure is called "Normal Tandem Solar Cell" (NTSC) (Fig. 3.3a).

The second model of tandem structure uses as front OPV cell active layer materials those mainly absorbing light of longer wavelengths and as back OPV cells active layer materials those mainly absorbing light of shorter wavelengths. This structure is called "Reverse Tandem Solar Cell" (RTSC) (Fig. 3.3b).

So, the interest of simulations will be focused on the study of the tow following structures:

- The device with P3HT: PCBM OPV cell as the front cell and pBBTDPP2: PCBM OPV cell as the back cell called NTSC.

The device with pBBTDPP2: PCBM OPV cell as the front cell and P3HT: PCBM OPV cell as the back cell called RTSC.

3.4.2 Simulation

The aim of the simulation is to optimize, in both configurations, the thicknesses of active layers of the two subcells verifying the condition of the current matching. Also, the current matching evolution as a function of the thicknesses of active layers of the two subcells is considered. Finally, a comparison between the optical characteristics of the two configurations is carried out in order to show the influence of the positions of the two subcells P3HT: PCBM and pBBTDPP2: PCBM with respect to each other on the optical characteristics of the tandem cell.

Like in the first section, the short-circuit current for the tow subcells is calculated assuming 100% internal quantum efficiency. For each of the two subcells, only the thickness of the blend active layer is chosen to be varied. The thicknesses of other layers correspond to those found after optimization in the previous section. The two structures are shown in Fig. 3.3. In both cases, the two subcells are separated by a ZnO layer of a thickness of 30 nm [6, 7] which acts as the recombination contact connecting the two subcells.

The simulation plan is presented as follows:

1. As a first step, 3D plots of J_{SC} Front and J_{SC} Back according to active layer thicknesses of the top and back cells d_{front} and d_{back} for the NTSC and RTSC configurations are calculated (Figs. 3.4a and 3.5a). The thicknesses of the active layers are varied between 0 and 200 nm with a calculation step of 5 nm.

2. The delimitation of the current matching as a function of the thicknesses of the active layers for the two configurations is also represented in Figs. 3.4b and 3.5b, respectively. The thicknesses of the active layers are varied between 0 and 300 nm with a calculation step of 5 nm.

3. The evolution of the current matching value as a function of the thicknesses of the active layers for the two configurations is presented in Figs. 3.4c and 3.5c active layer thicknesses are varied between 0 and 200 nm with a calculation step of 1 nm with an accuracy $\Delta J_{SC} = abs\ (J_{SC\_Top} - J_{SC\_Back}) \leq 0.05$ mA/cm^2

   From figures, it should be noted that:

   - In both NTSC and RTSC configurations (Figs. 3.4a and 3.5a), the top cell current is proportional to the increase in the thickness of its active layer and inversely proportional to the increase in the thickness of the back cell. However, the back cell current is proportional to the increase in the thickness of back cell active layer and inversely proportional to the increase of the top cell thickness.

   - For the NTSC configuration and for active layer thicknesses of the top cell above 225 nm (calculation thickness range between 0 and 300 nm), the top cell current is greater than the back cell current regardless of the thickness of the back cell active layer so that it is no longer possible to find a current matching between the two cells (Fig. 3.4b).

   - For the RTSC configuration, as soon as the thickness of the top cell active layer exceeds 105 nm (calculation thickness range between 0 and 300 nm), the top cell current becomes larger than the back cell current, regardless of the thickness of the back cell active layer so that it is no longer possible to find a current matching between the two cells (Fig. 3.5b).

   - In the NTSC configuration, the delimitation of current matching increases with the increase of the thickness of the top cell active layer, in this case the P3HT: PCBM blend layer (Fig. 3.4b). The current matching value also increases in proportion to the increase in thickness of the P3HT: PCBM layer to reach a maximum of 10.6945 mA/cm^2 for the thicknesses: 167 nm for the

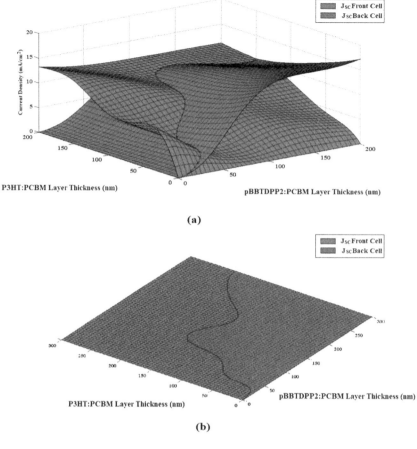

(a)

(b)

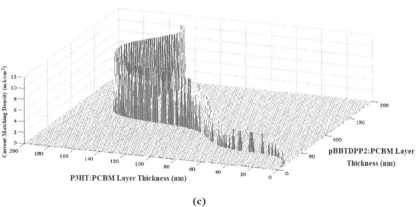

(c)

Fig. 3.4 For NTSC configuration: **a** 3D plots of J_{SC} Front and J_{SC} Back versus d_{front} and d_{back}. **b** Delimitation of the current matching as a function of d_{front} and d_{back}. **c** Evolution of the current matching value as a function of d_{front} and d_{back}

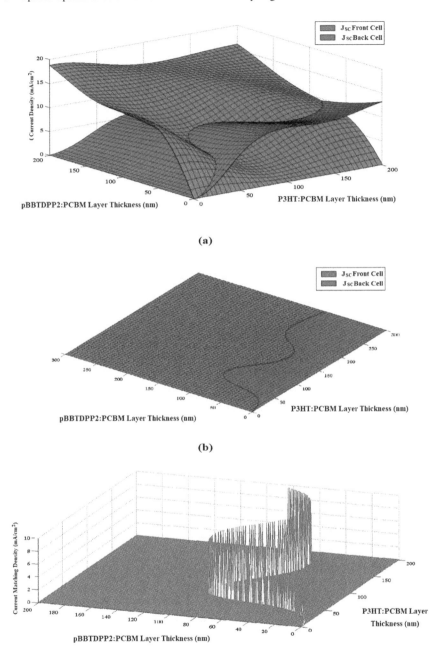

Fig. 3.5 For RTSC configuration: **a** 3D plots of J_{SC} Front and J_{SC} Back versus d_{front} and d_{back}. **b** Delimitation of the current matching as a function of d_{front} and d_{back}. **c** Evolution of the current matching value as a function of d_{front} and d_{back}

P3HT: PCBM blend layer and 131 nm for the pBBTDPP2: PCBM blend layer (Fig. 3.4c).

- In the RTSC configuration, the current matching is limited by a small thickness of the top cell active layer, in this case the pBBTDPP2: PCBM layer (Fig. 3.5b). Its thickness is limited to about 100 nm. This configuration provides relatively lower current matching values than those proposed by the NTSC configuration. The maximum value of the current matching, in this case 9.5557 mA/cm^2, is obtained for the thicknesses: 73 nm for the pBBTDPP2: PCBM active layer and 193 nm for the P3HT: PCBM active layer (Fig. 3.5c).

3.4.3 Results and Discussion

Before explaining these findings, let us first explain the following points:

1. For both NTSC and RTSC configurations:

   - The top cell is exposed directly to the light spectrum, while the back cell receives the light transmitted by the top cell after reflection and absorption of a part of the light spectrum by the latter (according to the values of the refractive index and extinction coefficient of its active layer). This contributes to weakening the current produced by the back cell.
   - When the thickness of the top cell active layer increases, there is more light absorption in the top cell, so its current will increase. At the same time, increasing the top cell active layer thickness will decrease the part of light transmitted to the lower cell, and its current will also decrease.
   - Increasing the thickness of the active layer of the bottom cell will increase its current due to the increase in light absorption. But the greater the thickness of the back cell active layer, the more the cathode Al is away from the top cell, minus the light fraction reflected by the latter to be absorbed a second time by the top cell. So, the top cell current is inversely proportional to the increase in the thickness of back cell active layer.

2. The pBBTDPP2: PCBM blend layer presents better optical characteristics than that offered by P3HT: PCBM blend layer over the entire visible spectrum. Indeed, the average value of the refractive index $n = 1.815154$ for PBBTDPP2: PCBM layer is better than the average value of $n = 1.896896$ for P3HT: PCBM layer. Also, the average value of the coefficient of extinction $k = 0.206749$ for PBBT-DPP2: PCBM layer is better than the average value of $k = 0.184112$ for the P3HT: PCBM layer (Table 3.2). As a result, we can conclude that:

   - For equal thicknesses, the pBBTDPP2: PCBM layer absorbs more light than the P3HT: PCBM layer.

- The amount of light transmitted to the pBBTDPP2: PCBM layer (after reflection at its interface with the PEDOT: PSS layer) is greater than that transmitted to the P3HT: PCBM layer (after reflection at its interface with the PEDOT: PSS layer).
- The pBBTDPP2: PCBM layer absorbs light mainly in the range of 600–800 nm but also has a significant absorption over the range of 350–600 nm, while the P3HT: PCBM blend layer only absorbs on the range of 300–650 nm (zero absorption over the range of 650–800 nm).

Based on the explanation provided above, the simulation results can be interpreted as follows:

- The P3HT: PCBM OPV cell placed above the tandem structure (NTSC) will see its optical characteristics improved in proportion to the increase in the thickness of its active layer since it is directly exposed to light. However, the part of the light spectrum between 600 nm and 800 nm (pBBTDPP2: PCBM absorption range) will be transmitted to the back cell without being absorbed by the top cell. As a result and taking into account the relatively high value of the extinction coefficient of the pBBTDPP2: PCBM layer, the back cell current is proportional to the increase in its active layer thickness and is not affected by the increase of the top cell active layer thickness in such way that for each thickness of the top cell, a thickness of the back cell giving a current matching can be found. However, by exceeding a threshold value of top cell active layer thickness $d_{front} = 225$ nm, it is no longer possible to find a thickness of the back cell active layer (calculation thickness range between 0 and 300 nm) giving a current matching (Fig. 3.4b). This can be explained by the fact that for these thicknesses, the fraction of light transmitted to the back cell is no longer sufficient to generate a current equal to that generated by the top cell regardless of the thickness of the back cell active layer. This explains results presented in the figure.
- The pBBTDPP2: PCBM OPV cell placed above the tandem structure (RTSC) will present high optical characteristics for low thicknesses of its active layer. The lower cell (P3HT: PCBM OPV cell), meanwhile, will see its optical characteristics diminished compared to that of the top cell since, in addition to the reasons cited above, the solar spectrum will be partially reflected and absorbed by the top cell before attacking the back cell. So, for top active layer thicknesses greater than 105 nm, the top cell current will remain higher than the back cell current regardless of the thickness of the active layer of the latter (calculation thickness range between 0 and 300 nm). Therefore, the current matching cannot be found for the tandem structure. This explains results presented in the figure (Fig. 3.5b).

In summary, simulation results show that the NTSC structure is more efficient for large thicknesses of the top cell active layer, whereas the RTSC structure is more efficient for thin thicknesses of this layer. This is in harmony with the literature [7].

Tables 3.3 and 3.4 present the current matching calculated with a precision of $\Delta J_{SC} \leq 0.01$ mA/cm^2 with the corresponding thicknesses of the active layers d_{Top} and d_{Back} for the two configurations NTSC and RTSC, respectively. It is easy to see from these tables that the NTSC configuration is more efficient for thick top cells, while the RTSC configuration is more efficient for fine top cells.

Table 3.3 gives, for the NTSC configuration, a maximum current value of 10.674627 mA/cm^2 for the thicknesses $d_{Top} = 167$ nm and $d_{Back} = 133$ nm, and for the RTSC configuration, a maximum current value of 9.546354 mA/cm^2 is obtained for the thicknesses $d_{Top} = 72$ nm and $d_{Back} = 192$ nm (Table 3.4).

Table 3.3 Current matching points for the NTSC configuration in a range of active layer thicknesses of between 0–200 nm and $\Delta J_{SC} = $ abs $(J_{SC\_Top} - J_{SC\_Back}) \leq 0.01$ mA/cm^2

$D_{Top \, (nm)}$	$d_{Back \, (nm)}$	$J_{SC\_Top \, (mA/cm^2)}$	$J_{SC\_Back \, (mA/cm^2)}$	$\Delta J_{SC \, (mA/cm^2)}$
2	16	0.859025	0.853596	0.005429
21	37	3.527084	3.533429	0.006345
36	39	3.563764	3.557578	0.006186
45	40	3.529252	3.526965	0.002287
59	44	3.849023	3.843462	0.005561
64	47	4.204539	4.196818	0.007721
76	60	5.996526	6.001096	0.00457
79	64	6.588809	6.580282	0.008527
82	68	7.156796	7.155384	0.001412
91	77	8.376469	8.368018	0.008451
136	90	9.645963	9.648004	0.002041
145	175	10.360377	10.359192	0.001185
146	171	10.402372	10.408182	0.00581
150	162	10.506447	10.508362	0.001915
151	160	10.52563	10.53159	0.00596
153	157	10.562101	10.554645	0.007456
156	152	10.60338	10.600635	0.002745
159	147	10.632748	10.640298	0.00755
160	98	10.049382	10.04182	0.007562
161	144	10.650491	10.653257	0.002766
163	141	10.66581	10.660243	0.005567
165	137	10.668696	10.675583	0.006887
166	135	10.669505	10.677374	0.007869
167	**133**	**10.670155**	**10.674627**	**0.004472**
168	131	10.670836	10.666866	0.00397
169	128	10.659894	10.658737	0.001157
170	124	10.637069	10.635842	0.001227

Table 3.4 Current matching points for RTSC configuration in a range of active layer thicknesses of between 0–200 nm and $\Delta J_{SC} = abs (J_{SC\_Top} - J_{SC\_Back}) \leq 0.01$ mA/cm^2

d_{Top} (nm)	d_{Back} (nm)	J_{SC_Top} (mA/cm^2)	J_{SC_Back} (mA/cm^2)	ΔJ_{SC} (mA/cm^2)
1	7	0.354480	0.352030	0.002450
6	22	2.088690	2.096564	0.007874
26	43	5.231499	5.228532	0.002967
28	44	5.333642	5.324951	0.008691
47	50	5.716535	5.724790	0.008255
56	155	8.664788	8.666537	0.001749
58	52	5.827432	5.820699	0.006733
59	177	9.296125	9.305737	0.009612
62	182	9.416800	9.426582	0.009782
64	135	8.340595	8.340750	0.000155
66	132	8.333205	8.324846	0.008359
68	129	8.314619	8.322295	0.007676
71	125	8.316731	8.324639	0.007908
72	**192**	**9.537747**	**9.546354**	**0.008607**
75	120	8.334149	8.338770	0.004621
79	197	9.512983	9.522675	0.009692
80	114	8.370322	8.360650	0.009672
84	109	8.385575	8.377437	0.008138
87	105	8.384643	8.382531	0.002112
90	63	6.889741	6.886543	0.003198
94	67	7.223257	7.214953	0.008304
98	78	7.871290	7.867440	0.003850

So, the NTSC configuration is more efficient than the RTSC one. This is probably due to the fact that for the NTSC configuration, the lowest efficiency cell is placed at the top of the tandem structure, which makes possible to better exploit its optical properties. The back cell absorbs essentially the part of the light between 600 nm and 800 nm. Therefore, it is little influenced by the top cell (apart from the part of the light reflected by the latter).

The reverse (RTSC) configuration presents a disadvantage of the low efficiency cell placed at the back position. So, the non-zero absorption profile of the top cell in the wavelength range of 350–600 nm in addition to the reflection phenomenon contributes to decrease the current produced by the back cell.

Figure 3.6 presents the calculated distributions of the normalized modulus squared of the optical electric field $|E|^2$ inside (a) 167–133 nm (optimized) NTSC. (c) 72–192 nm (optimized) RTSC and the fraction of light absorbed for (b) 167–133 nm NTSC. (d) 72–192 nm RTSC. For calculating the intensity of the electromagnetic field in the device, two wavelengths have been selected; 500 nm and 727 nm. The

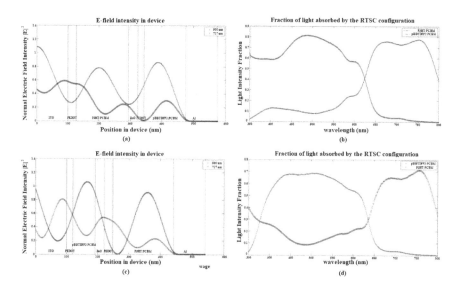

Fig. 3.6 **a** and **c** Calculated distributions of the normalized modulus squared of the optical electric field $|E|^2$ inside 167–133 nm NTSC and 72–192 nm RTSC configurations, respectively (for 500 and 727 nm wavelengths). **b** and **d** Fraction of light absorbed for 167–133 nm NTSC and 72–192 nm RTSC, respectively

first corresponds to the absorption peak of the P3HT: PCBM blend layer, and the second corresponds to the absorption peak of the pBBTDPP2: PCBM blend layer.

It should be noted that for the two wavelengths 500 nm and 727 nm, the E field shows a good distribution in the corresponding active layers in both configurations (NTSC and RTSC). For the 500 nm wavelength, the E field has a good distribution and is well absorbed in the P3HT: PCBM blend layer (Fig. 3.6a, c). The same remarks can be made for the 727 nm wavelength in both configurations. However, the 727 nm wave is more absorbed in the NTSC configuration ($|E|^2$ normalized value <1 in the pBBTDPP2: PCBM layer) than in the RTSC ($|E|^2$ normalized value >1 in the pBBTDPP2: PCBM layer (top cell) because of the interferences caused by the wave reflected by the cathode Al (and transmitted by the back cell). Indeed, because of the small thickness of the pBBTDPP2: PCBM active layer in the RTSC configuration, there is not enough material to properly absorb the 727 nm wavelength. In addition, the active layer P3HT: PCBM is transparent for this wavelength. So, it will be reflected by the cathode Al toward the top cell what will amplify the wave by interference.

Figure 3.6b, d show the absorption rate of the tandem structures in NTSC and RTSC configurations, respectively. It is easy to see the advantage offered by the use of this type of structure since the absorption of the light is done over the entire spectrum of the visible. However, the NTSC structure presents a better absorption rate with peaks reaching 80% of the incident light intensity compared to 70% offered by the RTSC configuration.

3.5 Concluding Remarks

In this chapter, it was chosen to optimize the tandem structure composed of OPV cells based on interpenetrating materials P3HT: PCBM and pBBTDPP2: PCBM. At first, the optical yield of the OPV cell based on P3HT: PCBM interpenetrating blend was optimized (Sect. 3.1). The simulation results showed a good absorption of the electrical field (E) with a good distribution profile over the entire thickness of the active layer. The results were close to those available in the literature [8, 11]. After this, a comparative study of the optical efficiency of the pBBTDPP2: PCBM blend OPV cell with some OPV cells based on materials commonly used in the field of organic photovoltaics was carried out (Sect. 3.2). This study resulted in the possibility of using the pBBTDPP2: PCBM and P3HT: PCBM blend OPV cells in a tandem structure to widen the absorbed light range to the entire visible spectrum and thus to improve optical efficiency.

Finally, the study of tandem structure based on P3HT: PCBM and pBBTDPP2: PCBM blend materials was discussed. The arrangement of the two cells with respect to each other was taken into account (NTSC and RTSC configurations). The aim being to optimize the thickness of active layers of the two sub-cells verifying the condition of the current matching. The simulation results showed that the NTSC configuration is more efficient for the large thicknesses of the top cell, whereas the RTSC is more efficient for the thin thicknesses of the top cell which is in agreement with the literature [7]. Also, the NTSC configuration offers better optical characteristics than that proposed by the RTSC. The simulation results showed an absorption rate covering the entire spectrum of the visible, which presents a real advantage of using this type of structure in order to improve the yield of OPV cells.

At the end of this chapter, it can be said that optical modeling is an effective tool that guides the experimental in the interpretation of its measurements. It can even be essential in the design of tandem cell structures. However, given the variety of parameters on which the efficiency of organic photovoltaic cells depends, optical modeling will have to be followed by electrical modeling; this is to represent the steps describing the organic photovoltaic conversion process. This model considers the exciton generation (output of the optical simulation), recombination, drift, diffusion and collection process of the electron and hole in the active BHJ layer. This will refine the simulation results and make them closer to reality.

Acknowledgements The second author would like to thank the Simons Foundation for financial support. A part of this work was carried out at the ICTP, Trieste, Italy.

References

1. Dennler, G., Scharber, M.C., Brabec, C.J.: Polymer-fullerene bulk-heterojunction solar cells. Adv. Mater. **21**, 1323–1338 (2009)
2. Nelson, J.: Polymer: fullerene bulk heterojunction solar cells. Mater. Todays **14**(10), 462–470 (2011)

3. Gevaerts, V.S., Anton Koster, L.J., Wienk, M.M., Janssen, R.A.J.: Discriminating between bilayer and bulk heterojunction polymer: fullerene solar cells using the external quantum efficiency. ACS Appl. Mater. Interfaces. **3**, 3252–3255 (2011)
4. Rasi, D.D.C., Janssen, R.A.J.: Advances in solution-processed multijunction organic solar cells. Adv. Mater. **31**, 1806499 (2019)
5. Meng, L., Zhang, Y., Wan, X., Li, C., Zhang, X., Wang, Y., Ke, X., Xiao, Z., Ding, L., Xia, R., Yip, H.-L., Cao, Y. Chen, Y.: Organic and solution-processed tandem solar cells with 17.3% efficiency. Science **361**, 1094–1098 (2018)
6. Gilot, J., Wienk, M.M., Janssen, R.A.J.: Optimizing polymer tandem solar cells. Adv. Energy Mater. **22**, E67–E71 (2010)
7. Wang, Z., Zhang, C., Chen, D., Zhang, J., Feng, Q., Xu, S., Zhou, X., Hao, Y.: Investigation of controlled current matching in polymer tandem solar cells considering different layer sequences and optical spacer. Jpn. J. Appl. Phys. **51**, 122301 (2012)
8. Pettersson, L.A.A., Roman, L.S., Inganas, O.: Modeling photocurrent action spectra of photovoltaic devices based on organic thin films. J. Appl. Phys. **86**(1), 487–496 (1999)
9. Peumans, P, Yakimov, A., Forrest, S.R.: Small molecular weight organic thin-film photodetectors and solar cells. J. Appl. Phys. **93**(7) (2003)
10. Monestier, F., Simon, J.J., Torchio, P., Escoubas, L., Flory, F., Bailly, S., de Bettignies, R., Guillerez, S., Defranoux, C.: Modeling the short-circuit current density of polymer solar cells based on P3HT:P CBM blend. Solar Energy Mater. Solar Cells **91**(5), 405–410 (2007)
11. Nam, Y., Huh, J., Jo, W.H.: Optimization of thickness and morphology of active layer for high performance of bulk-heterojunction organic solar cells. Sol. Energy Mater. Sol. Cells **94**, 1118–1124 (2010)
12. Liang, C., Wang, Y., Li, D., Ji, X., Zhang, F., He, Z.: Modeling and simulation of bulk heterojunction polymer solar cells. Sol. Energy Mater. Sol. Cells **127**, 67–86 (2014)
13. Burkhard, G.F., Hoke, E.T.: Transfert Matrix Optical Modeling. McGehee Group (Stanford Univ). http://www.stanford.edu/group/mcgehee (2011)
14. Burkhard, G.F., Hoke, E.T., McGehee, M.D.: Accounting for interference, scattering, and electrode absorption to make accurate internal quantum efficiency measurements in organic and other thin solar cells. Adv. Mater. **22**, 3293–3297 (2010)
15. Vedraine, S., Torchio, Ph., Derbal-Habak, H., Flory, F., Brissonneau, V., Duché, D., Simon, J.J., Escoubas, L.: Plasmonic structures integrated in organic solar cells. Proc. SPIE Conf. **7772**, 777219-1–6777219-6 (2010)
16. Guedes, A.F.S., Guedes,, V.P., Tartari, S., Souza, M.L., Cunha, I.J.: New organic semiconductor materials applied in organic photovoltaic and optical devices. Systemics, Cybern. Inform. **13**(2), 38–40 (2015)
17. Abada, Z., Mellit, A.: Optical optimization of organic solar cells based on P3HT: PCBM interpenetrating blend. In: 2017 5th International Conference on Electrical Engineering—Boumerdes (ICEE-B), pp. 1–6, (2017). https://doi.org/10.1109/icee-b.2017.8191966
18. Shen, H., Maes, B.: Combined plasmonic gratings in organic. Opt. Express **19**(S6), A1202–A1210 (2011)
19. Niggemann, M., Ziegler, T., Glatthaar, M., Riede, M., Zimmermann, B., Gombert, A.: Optical near field phenomena in planar and structured organic solar cells. Proc. SPIE Photonics Sol. Energy Syst. **6197**, 61970D-1–61970D-10 (2006)
20. Armin, A., Velusamy, M., Wolfer, P., Zhang, Y., Burn, P.L., Meredith, P., Pivrikas, A.: Quantum efficiency of organic solar cells: electro-optical cavity considerations. ACS Photonics **1**(3), 173–181 (2014)
21. Supriyanto, A., Mustaqim, A., Agustin, M., Ramelan, A.H., Septa Rosa, S.E., Yofentina, Nurosyid, F.: Fabrication of organic solar cells with design blend P3HT: PCBM variation of mass ratio. In: IOP Conference Series: Materials Science and Engineering. vol. 107, p. 012050. https://doi.org/10.1088/1757-899x/107/1/012050 (2016)
22. You, J., Dou, L., Hong, Z., Li, G., Yang, Y.: Recent trends in polymer tandem solar cells research. Progress Polym. Sci. **38**, 1909–1928 (2013)

Chapter 4
Theoretical Study of Quantum Well GaAsP(N)/GaP Structures for Solar Cells

L. Chenini and A. Aissat

Abstract In this theoretical study, we simulate a compressively strained $GaAs_xP_{1-x}$ and tensile strain $GaN_yAs_xP_{1-x-y}$ quantum well active zones with the aim to be inserted in solar cells. We will compare the ternary GaAsP/GaP and quaternary GaNAsP/GaP quantum well structures (QWs) by modeling these two types of systems. We first studied the introduction effect of arsenic (As) into the host material GaP and the optoelectronic properties of the obtained ternary alloy GaAsP. Then, we study the additional effect of just a few percent of nitrogen (N) in this ternary alloy. Incorporate nitrogen into the GaAsP alloy has been shown to split the conduction band into two bands E_+ and E_- causing a reduction of the band gap energy. Addition of nitrogen to GaAsP induces a redshift in the emission wavelengths and increases the absorption coefficient.

Keywords Materials · Nanostructures · Quantum well · Optoelectronic · Photovoltaic

4.1 Introduction

Photovoltaic is a non-polluting, ecological, and powerful way to provide electricity and supply the growing global demand for energy. Indeed, it allows the use of a renewable energy source (the Sun) considered inexhaustible on a human scale and converts it directly into electrical energy. The intensive development of silicon semiconductor technology has predetermined their widespread distribution observed today. In industry, the basis of technology for the silicon purification, processing, and production has been at the highest level among all semiconductor materials by now and the integration of III–V semiconductors grown on Si substrates allows for significant cost reduction of photovoltaic with very high conversion efficiency

L. Chenini · A. Aissat (✉)
Faculty of Technology, University Blida 1, Blida, Algeria
e-mail: sakre23@yahoo.fr

© The Editor(s) (if applicable) and The Author(s), under exclusive license to Springer Nature Switzerland AG 2020
A. Mellit and M. Benghanem (eds.), *A Practical Guide for Advanced Methods in Solar Photovoltaic Systems*, Advanced Structured Materials 128,
https://doi.org/10.1007/978-3-030-43473-1_4

[1, 2]. It is very important to search for new semiconductor in order to ameliorate the solar cell properties [3, 4].

The dilute nitrides III–N–V form a new family of perspective alloys for optoelectronic devices such as lasers, light-emitting diodes, and high-efficiency solar cells [5–12]. The GaNAsP material is one of these classes of material. Quantum well solar cells consist of inserting a multiple quantum well system in the intrinsic region of a p-i-n cell in order to improve and adjust the spectral response of the cell [13–15]. A considerable number of theoretical and experimental studies have been made on GaAsP and GaNAsP grown on Si and GaP substrates [16–27]. Previously, we have studied the intersubband absorption coefficient in GaAsP/GaP quantum well and verified the dependence of this parameter on the arsenic concentration and quantum well width [28].

In this work, we have investigated compressively strained $GaAs_xP_{1-x}$ and tensile strained $GaN_yAs_xP_{1-x-y}$ quantum well active zones grown on a GaP substrate. These two structures are chosen to be inserted into solar cells in order to achieve high-performance quantum well solar cells, that is to say, having a growth of high-quality materials which ensures efficient and maximum light absorption. x and y are the arsenic and nitrogen molar fractions ranged, respectively, (between (0–0.4) and (0–0.05)).

Simulation is employed to calculate the material parameters and to be able to make a comparison between these two structures. First, we compare the lattice parameter, band gap energy, strain, and band offset ratios of these quantum well systems. The major importance of this approach is the simultaneous reduction of the lattice parameter and the band gap of the GaNAsP alloy. A very unusual property for band gap engineering in the semiconductor alloys we have accustomed is that the reduction of the band gap energy is generally followed by an increase in the lattice parameter. Then, the interband transition energy at the Γ point and the corresponding emission wavelength as functions of the QWs widths with arbitrary compositions of the two structures are established. Finally, the composition effects on absorption coefficient are also discussed.

4.2 Theoretical Model

Figure 4.1a, b illustrate the active regions of the two structures which will be studied and compared. The first one (a) consists of a thin GaAsP layer sandwiched between two layers of a gallium phosphide (GaP) material. In the second structure (b), a thin layer of GaNAsP will be inserted between the two barrier layers of GaP. To get the different parameters for both material systems $GaAs_xP_{1-x}$ and $GaN_yAs_xP_{1-x-y}$, we linearly interpolate the parameters of the binary semiconductors which can be found in Table 4.1 [29], except for the unstrained band gap energies. The interpolation formulas will be given as [30]:

$$\Theta(GaAs_xP_{1-x}) = x.\,\Theta(GaAs) + (1-x).\,\Theta(GaP) \qquad (4.1)$$

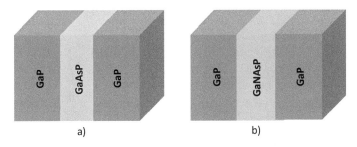

Fig. 4.1 GaAsP/GaP and GaNAsP/GaPstructures

Table 4.1 Parameters used in this work at 300 K

Parameters	Symbol	GaN	GaP	GaAs
Elastic constant	C_{11} (10^{11} dyn cm^{-2})	29.3	14.05	11.88
Elastic constant	C_{12} (10^{11} dyn cm^{-2})	15.9	6.203	5.38
Spin-orbit energy	Δ_{SO} (eV)	0.017	0.08	0.34
Shear deformation potential	b (eV)	-2.2	-1.7	-1.85
Hydrostatic deformation potential for CB	a_c (eV)	-2.2	-7.14	-11
Hydrostatic deformation potential for VB	a_v (eV)	5.2	1.70	-0.85
Electron effective mass	m_e/m_0	0.21	0.114	0.067
Heavy hole effective mass	m_{hh}/m_0	1.27	0.52	0.55
Band gap energy	E_g (eV)	3.42	2.80	1.42
Lattice parameter	a (Å)	4.52	5.4508	5.6533

$$\Theta\left(GaN_yAs_xP_{1-x-y}\right) = x.\Theta(GaAs) + y.\Theta(GaN)$$
$$+ (1 - x - y).\Theta(GaP) \qquad (4.2)$$

The band gap energy for GaAs$_x$P$_{1-x}$ alloy is obtained by the following expression:

$$E_g(GaAs_xP_{1-x}) = x.E_g(GaAs)$$
$$+ (1 - x).E_g(GaP) - 0.19.x.(1 - x) \qquad (4.3)$$

For GaNAsP alloy, the band gap is obtained using the anticrossing model [31, 32]:

$$E_g(GaNAsP) = \frac{1}{2}\Big[E_g(GaAsP) + E_N$$
$$- \sqrt{\left(E_g(GaAsP) - E_N\right)^2 + 4C_{MN}^2 y}\Big] \qquad (4.4)$$

The anticrossing BAC model has been agreeably used to explain the dependences of the lower E_- and upper E_+ subband energies on the nitrogen concentration. This model is explained as the interaction of the highly localized energy level E_N induced by N atoms and the lowest conduction band of the GaAsP semiconductor matrix. C_{MN} is the interband matrix element; these two parameters are obtained by interpolation, as explained above; and the parameters of GaNAs [33] and GaAsP [34] are, respectively, equal to $E_N^{\text{GaNAs}} = 1.65\,\text{eV}$, $C_{MN}^{\text{GaNAs}} = 2.7\,\text{eV}$ and $E_N^{\text{GaAsP}} = 2.25\,\text{eV}$, $C_{MN}^{\text{GaAsP}} = 3.16\,\text{eV}$.

The strain in both structures has been calculated and has the following components:

$$\varepsilon_{xx}(\text{GaAsP, GaNAsP}) = \frac{a_0(\text{GaP})}{a(\text{GaAsP, GaNAsP})} - 1 \tag{4.5}$$

$$\varepsilon_{zz}(\text{GaAsP, GaNAsP}) = -2\frac{C_{12}}{C_{11}}.\varepsilon_{xx}(\text{GaAsP, GaNAsP}) \tag{4.6}$$

where a (GaAsP, GaNAsP) is the lattice constant of the two quantum well structures, a_0 is the lattice constant of the substrate GaP, and C_{11} and C_{12} are the elastic stiffness constants.

The band offset ratios of the conduction and valence band of both alloys can be given by:

$$Q_c = \frac{E_c^w - E_c^b}{E_g^w - E_g^b} \tag{4.7}$$

$$Q_v = 1 - Q_c \tag{4.8}$$

where E_c^w and E_c^b are, respectively, the conduction band position in the quantum well and the barrier materials, E_g^w and E_g^b correspond, respectively, to the strain band gaps for the quantum well and barrier alloys. The theoretical model to obtain E_c^w and E_c^b is well explained in reference [30]. The emission wavelength is mainly determined after calculating the transitns, E_{tra}. The photon energy of transition is expressed as [30]:

$$\lambda = \frac{1.24}{E_{\text{tra}}} \tag{4.9}$$

The transition energy E_{tra} is defined as the transitions between the first subband holes in the valence band and the first subband electrons in the conduction band.

All incident photons with energy (E_{ph}) equal to or greater than the band gap energy (E_g) will be absorbed. The absorption coefficient was obtained with the help of the following expression:

$$\alpha = \alpha_0 \frac{\sqrt{E_{\text{ph}} - E_g}}{E_{\text{ph}}} \tag{4.10}$$

α_0 is a constant depending on the density of states related to the absorption of photons.

4.3 Results and Discussion

Introducing arsenic in the host material (GaP) leads to the reduction of the band gap energy as we can see in Fig. 4.2. Arsenic is an atom with covalent radius and electronegativity of 1.19 Å and 2.18, respectively, while nitrogen has a smaller radius of 0.75 Å but has a high electronegativity with a value of 3.04. These differences in size and electronegativity create a perturbation that will lead to the splitting of the conduction band into two subbands E_+ and E_- as illustrated in Fig. 4.3.

The higher energy E_+ increases with nitrogen concentration and decreases with arsenic concentration. On the other hand, the lower energy E_- shifts strongly toward lower values with an increase in nitrogen concentration but decreases slightly with increasing arsenic concentration. We have verified the change in lattice parameter with increasing arsenic concentration for GaAs$_x$P$_{1-x}$ on the GaP substrate using Eq. 4.1 as we can see in Fig. 4.4. It is clear that adding As leads to an increase in the lattice parameter. Figure 4.5 shows the calculated values of lattice parameter with increasing both arsenic and nitrogen concentration for GaN$_y$As$_x$P$_{1-x-y}$/GaP structure using Eq. 4.2. As can be seen from this figure, the lattice parameter increases by adding arsenic concentration to the system, but decreases with introducing nitrogen.

Increasing the arsenic concentration leads to a decrease in the strain in the (x, y) plane. It is found that the material undergoes compressive strain as shown in Fig. 4.6.

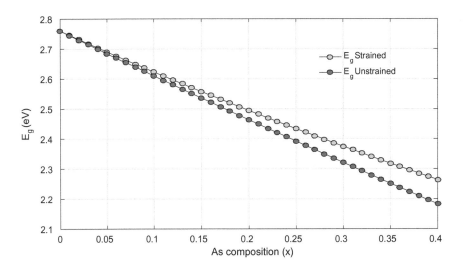

Fig. 4.2 Strained and unstrained band gap energy of GaAsP material as a function of As concentration

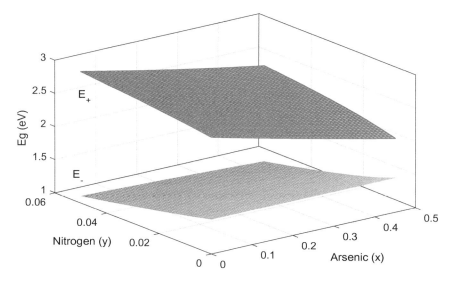

Fig. 4.3 Band gap energy of GaNAsP material as a function of As and N concentrations

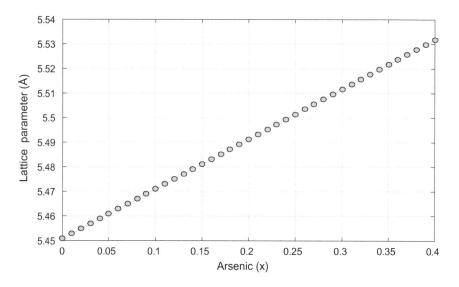

Fig. 4.4 Lattice parameter of GaAsP material as a function of As concentration

Figure 4.7 depicts the calculated results of the strain of the $GaN_yAs_xP_{1-x-y}$ well with GaP barriers. This figure reveals that strain in this structure increases with increasing nitrogen and arsenic concentration. Area of positive values represents the couples (x, y) which lead to a tensile strain, the negative values area represents the

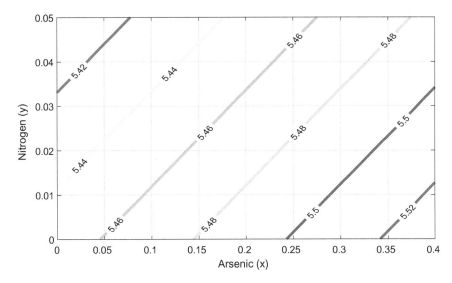

Fig. 4.5 Lattice parameter of GaNAsP QWs as a function of As and N concentrations

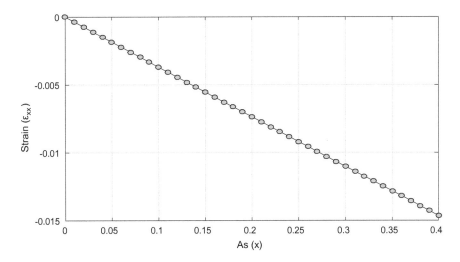

Fig. 4.6 Variation of strain versus arsenic concentration of GaAsP/GaP

couples which lead to a compressive strain, and the line of "0" represents the values for which GaNAsP can be matched to GaP.

In Fig. 4.8, we present the arsenic fraction molar dependence of conduction (Q_c) and valence band (Q_v) offset ratios of GaAsP well material grown on a GaP substrate. The addition of As to GaP causes changes in the band alignments. This figure illustrates the fact that the introduction of arsenic into GaP decreases Q_c whereas it increases the Q_v but Q_c takes values greater than that of the Q_v.

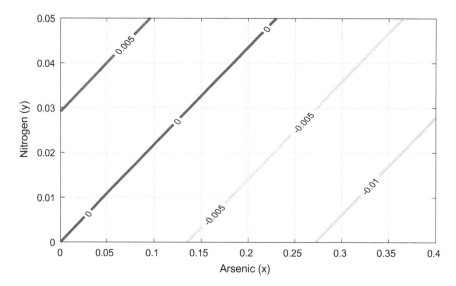

Fig. 4.7 Two-dimensional representation of the variation of strain versus arsenic and nitrogen concentration of the GaNAsP/GaP structure system

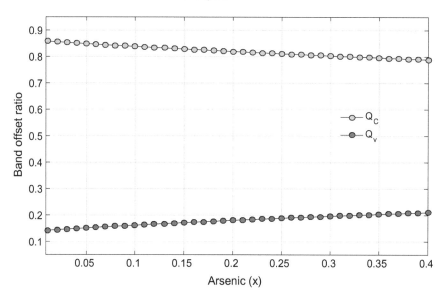

Fig. 4.8 The arsenic As dependence of both: conduction band offset ratio Q_c and valence band offset ratio Q_v of GaAs$_x$P$_{1-x}$/GaP QWs

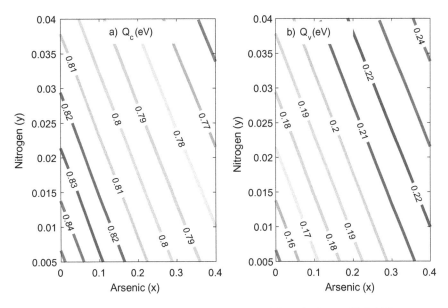

Fig. 4.9 Arsenic and nitrogen dependence of Q_c and Q_v of GaN$_y$As$_x$P$_{1-x-y}$/GaP QWs

We have presented the variations of calculated conduction (Q_c) and valence band (Q_v) offset ratios of GaN$_y$As$_x$P$_{1-x-y}$ quantum well grown on a GaP substrate in Fig. 4.9a, b. The addition of N to GaAsP causes substantial changes in the offset ratios; increasing the amount quantities of N into GaAsP decreases the conduction band offset ratio Q_c and increases the valence band offset ratio Q_v, see Fig. 4.9a, b. We can have deeper wells, leading to much better confinement of carriers.

Figure 4.10a plots the transition energy of the compressively GaAsP/GaP QWs as a function of well width for two values of arsenic ($x = 0.15$ and $x = 0.35$) while Fig. 4.10b gives the corresponding wavelength emission. Increasing the quantum well width decreases the transition energy and increasing arsenic redshift the corresponding wavelength emission. Figure 4.11a depicts the transition energy of the tensile GaN$_{0.005}$As$_x$P$_{0.995-x}$/GaP QWs as a function of well width for two values of arsenic ($x = 0.15$ and $x = 0.35$) at room temperature, Fig. 4.11b illustrates the corresponding wavelength emission. We can see that for values of well width less than 12 Å, both structures GaN$_{0.005}$As$_{0.15}$P$_{0.845}$/GaP (blue line) and GaN$_{0.005}$As$_{0.35}$P$_{0.645}$/GaP (red line) will have approximately the same transition energy value, but when the well width exceeds this value, increasing the arsenic concentration leads to decrease the transition energy which means redshift the emission wavelength. When the well thickness exceeds 70 Å, the transition energies will saturate and will have the same values of about 1.621 eV ($\lambda = 0.765$ μm) for $x = 0.15$ and 1.524 eV ($\lambda = 0.813$ μm) for $x = 0.35$, and in these cases, the quantum confinement is neglectable.

Figure 4.12a shows the dependence of the interband transition energy at $T = 300$ K in the tensile GaN$_{0.02}$As$_x$P$_{0.98-x}$/GaP QWs for the two values of arsenic ($x =$

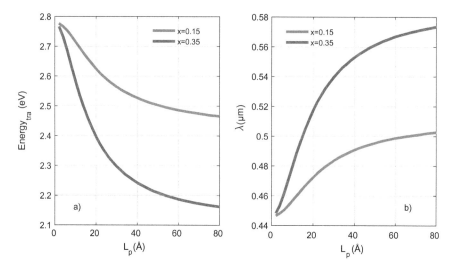

Fig. 4.10 Dependence of **a** transition energy and **b** emission wavelength of GaAs$_x$P$_{1-x}$/GaP ($x = 0.15, x = 0.35$) QWs on well width

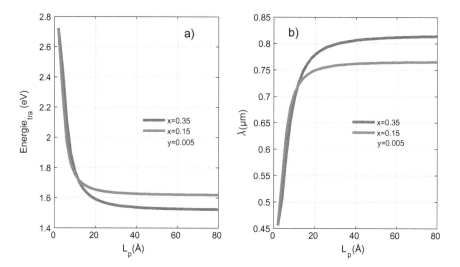

Fig. 4.11 Dependence of **a** transition energy and **b** emission wavelength of GaN$_{0.005}$As$_x$P$_{0.995-x}$/GaP ($x = 0.15, x = 0.35$) QWs on well width

0.15 and $x = 0.35$). Figure 4.12b illustrates the corresponding wavelength emission. When the nitrogen concentration rises, the transition energy decreases which leads to redshift the emission wavelength. As in Fig. 4.11, when the well width exceeds 70 Å, the transition energy becomes to be the same value of about 1.521 eV ($\lambda = 0.815$ μm) for $x = 0.15$ and 1.418 eV ($\lambda = 0.874$ μm) for $x = 0.35$.

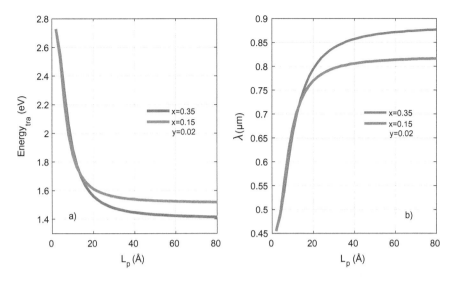

Fig. 4.12 Dependence of **a** transition energy and **b** emission wavelength of GaN$_{0.02}$As$_x$P$_{0.98-x}$/GaP ($x = 0.15$, $x = 0.35$) QWs on well width

Figures 4.13 and 4.14 illustrate, respectively, the dependence of the absorption coefficient on photon energy for several arsenic concentrations of GaAs$_x$P$_{1-x}$/GaP QWs and several nitrogen concentrations of GaN$_y$As$_{0.35}$P$_{0.65-x}$/GaP QWs. Addition of arsenic in host material increases the absorption coefficient and slightly redshift the corresponding wavelength. Increasing the fraction of included nitrogen just in few quantities in the GaAsP material is associated with the strong redshift of the wavelength and increases the absorption coefficient values.

4.4 Conclusion

In this chapter, we have evaluated the potential of ternary and quaternary alloys GaAsP and GaAsPN grown on the GaP substrate, for the development of a solar cell. The present study demonstrates that the addition of nitrogen into the GaAsP/GaP QWs has remarkable properties. It leads to split the conduction band into two subbands which contribute to the reduction of the band gap energy. Our numerical simulation indicates that we will have a considerable increase in the absorption coefficient and redshift of the corresponding wavelength. These two material systems are promising candidates for future quantum well solar cells.

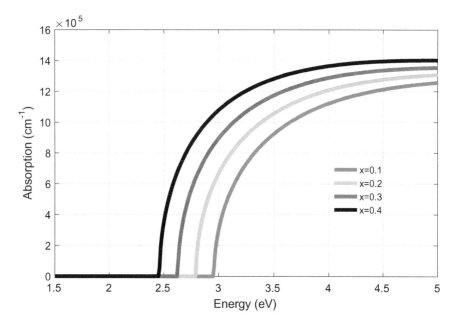

Fig. 4.13 Dependence of absorption coefficient on photon energy for several arsenic concentrations of GaAs$_x$P$_{1-x}$/GaP QWs

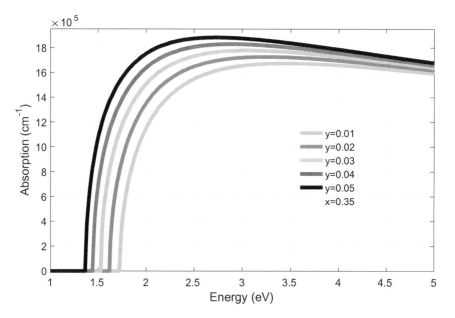

Fig. 4.14 Dependence of absorption coefficient on photon energy for several nitrogen concentrations of GaN$_y$As$_{0.35}$P$_{0.65-x}$/GaP QWs

References

1. Tibbits, T.N.D., Beutel, P., Grave, M., Karcher, C., Oliva, E., Siefer, G. et al.: New efficiency frontiers with wafer-bonded multi-junction solar cells. In: Proceeding, The 29th European PV Solar Energy Conference and Exhibition. pp. 1–4 (2014)
2. Durand, O., Almosni, S., Cornet, C., Létoublon, A., Levallois, C., Rolland, A., Even, J., Rale, P., Lombez, L., Guillemoles, J.F.: Multijunctionphotovoltavics: integrating III–V semiconductor heterostructures on silicon. SPIE 1–3 (2015)
3. Luceño-Sánchez, J.A., Díez-Pascual, A.M., Capilla, R.P.: Materials for photovoltaics: state of art and recent developments. Int. J. Mol. Sci. **20**, 1–42 (2019)
4. Almosni, S., Delamarre, A., Jehla, Z., Suche, D. et al.: Material challenges for solar cells in the twenty-first century: directions in emerging technologies. Sci. Technol. Adv. Mater. **19**, 336–369 (2018)
5. Mawst, L.J., Kim, T.W., Kim, H., Kim, Y., Kim, K., Lee, J.J., Kuech, T.F., Lingley, Z.R., LaLumondiere, S.D., Sin, Y., Lotshaw, W.T., Moss, S.C.: Dilute-nitride-antimonide materials grown by MOVPE for multi-junction solar cell application. ECS Trans. **66**, 101–108 (2015)
6. Kim, T.W., Forghani, K., Mawst, L.J., Kuech, T.F., LaLumondiere, S.D., Sin, Y., Lotshaw, W.T., Moss, S.C.: Properties of 'bulk' GaAsSbN/GaAs for multi-junction solar cell application: reduction of carbon background concentration. J. Cryst. Growth **393**, 70–74 (2014)
7. Kim, T.W., Kuech, T.F., Mawst, L.J.: Impact of growth temperature and substrate orientation on dilute-nitride-antimonide materials grown by MOVPE for multi-junction solar cell application. J. Cryst. Growth **405**, 87–91 (2014)
8. Gubanov, A., Polojärvi, V., Aho, A., Tukiainen, A., Tkachenko, N.V., Guina, M.: Dynamics of time-resolved photoluminescence in GaInNAs and GaNAsSb solar cells. Nanoscale Res. Lett. **9**, 1–4 (2014)
9. Tukiainen, A., Aho, A., Polojärvi, V., Ahorinta, R., Guina, M.: High efficiency dilute nitride solar cells: Simulations meet experiments. J. Green Eng. **5**, 113–132 (2016)
10. Rolland, A., Pedesseau, L., Even, J., Almosni, S., Robert, C., Cornet, C., Jancu, J.M., Benhlal, J., Durand, O., Le Corre, A., Rale, P., Lombez, L., Guillemoles, J.F., Tea, E., Laribi, S.: Design of a lattice-matched III–V–N/Si photovoltaic tandem cell monolithically integrated on silicon substrate. Opt. Quant. Electron. **46**, 1397–1403 (2014)
11. Zou, Y., Zhang, C., Honsberg, C., Vasileska, D., King, R., Goodnick, S.: A lattice-matched GaNP/Si three-terminal tandem solar cell. In: IEEE 7th World Conference on Photovoltaic Energy Conversion (WCPEC). pp. 279–282 (2018)
12. Durand, O., Almosni, S., Wang, Y.P., Cornet, C., Létoublon, A., et al.: Monolithic integration of diluted-nitride III–V–N compounds on silicon substrates: toward the III–V/Si concentrated photovoltaics. Energy Harvesting Syst. **1**, 147–156 (2014)
13. Bamham, K., Ballard, I., Barnes, J., Connolly, J., Griffin, P., Kluftinger, B., Nelson, J., Tsui, E., Zachariou, A.: Quantum well solar cells. Appl. Surf. Sci. **113**(114), 722–733 (1997)
14. Sayed, I., Bedair, S.M.: Quantum well solar cells: principles, recent progress, and potential. IEEE J. Photovoltaics **9**, 402–423 (2019)
15. Cabrera, C.I., Rimada, J.C., Courel, M., Hernandez, L., Connolly, J.P., Enciso, A., Contreras-Solorio, D.A.: Modeling multiple quantum well and superlattice solar cells. Nat. Resour. **4**, 235–245 (2013)
16. Vaisman, M., Fan, S., Yaung, K.N., Perl, E., Martín-Martín, D., Yu, Z.J., Leilaeioun, M., Holman, Z.C., Lee, M.L.: 15.3%-Efficient GaAsP solar cells on GaP/Si templates. ACS Energy Lett. **2**, 1–33 (2017)
17. Grassman, T.J., Chmielewski, D.J., Carnevale, S.D., Carlin, J.A., Ringel, S.A.: GaAs$_{0.75}$P$_{0.25}$/Si dual-junction solar cells grown by MBE and MOCVD. IEEE J. Photovoltaics **6**, 326–331 (2016)
18. Kim, B., Toprasertpong, K., Paszuk, A., Supplie, O., Nakano, Y., Hannappel, T., Sugiyama, M.: GaAsP/Si tandem solar cells: realistic prediction of efficiency gain by applying strain-balanced multiple quantum wells. Sol. Energy Mater. Sol. Cells **180**, 303–310 (2017)
19. Yaung, K.N., Lang, J.R., Lee, M.L.: Towards high efficiency GaAsP solar cells on (001) GaP/Si. In: IEEE 40th Photovoltaic Specialist Conference (PVSC). pp 831–835 (2014)

20. Fan, S., Yu, Z.J., Sun, Y., Weigand, W., Dhingra, P., Kim, M., Hoola, R.D., Ratta, E.D., Holman, Z.C., Lee, M.L.: 20% efficient epitaxial GaAsP/Si tandem solar cells. Sol. Energy Mater. Sol. Cells **202**, 110144 (1/8)

21. Shakfa, M.K., Woscholski, R., Gies, S., Wegele, T., Wiemer, M., Ludewig, P., Jandieri, K., Baranovskii, S.D., Stolz, W., Volz, K., Heimbrodt, W., Koch, M.: Carrier dynamics in Ga(NAsP)/Si multi-quantum well heterostructures with varying well thickness. Superlattices Microstruct. **93**, 67–72 (2016)

22. Almosni, S., Robert, C., Thanh, T.N., Cornet, C., Létoublon, A. et al.: Evaluation of InGaPN and GaAsPN materials lattice-matched to Si for multi-junction solar cells. J. Appl. Phys. **113**, 123509 (1/6) (2013)

23. Robert, C., Bondi, A., Nguyen Thanh, T., Even, J., Cornet, C. et al.: Room temperature operation of GaAsP(N)/GaP(N) quantum well based light-emitting diodes: effect of the incorporation of nitrogen. Appl. Phys. Lett. **98**, 251110 (2011)

24. Rosemann, N.W., Metzger, B., Kunert, B., Volz, K., Stolz, W., Chatterjee, S.: Temperature-dependent quantum efficiency of Ga(N, As, P) quantum wells. Appl. Phys. Lett. **103**, 252105 (2013)

25. Németh, I., Torunski, T., Kunert, B., Stolz, W., Volz, K.: Microstructural analysis of Ga(NAs)/GaP heterostructures. J. Appl. Phys. **101**, 123524 (2007)

26. Almosni, S., Rale, P., Cornet, C., Perrin, M., Lombez, L., Létoublon, A., Tavernier, K., Levallois, C., Rohel, T., Bertru, N., Guillemoles, J.F., Durand, O.: Correlations between electrical and optical properties in lattice-matched GaAsPN/GaP solar cells. Sol. Energy Mater. Sol. Cells **147**, 53–60 (2016)

27. Ilahi, S., Almosni, S., Chouchane, F., Perrin, M., Zelazna, K., Yacoubi, N., Kudrawiec, R., Râle, P., Lombez, L., Guillemoles, J.-F., Durand, O., Cornet, C.: Optical absorption and thermal conductivity of GaAsPN absorbers grown on GaP in view of their use in multijunction solar cells. Sol. Energy Mater. Sol. Cells **141**, 291–298 (2015)

28. Chenini, L., Aissat, A., Vilcot, J.P.: Theoretical study of intersubband absorption coefficient of conduction band in GaAsP/GaP quantum well structures. In: IEEE, 2019 International Conference on Power Generation Systems and Renewable Energy Technologies (PGSRET) (2019)

29. Adachi, S.: Material for optoelectronics and photonics. In: Springer Handbook of Electronic and Photonic Materials. pp. 725–741 (2017)

30. Zhang, Y., Ning, Y., Zhang, L., Zhang, J., Zhang, J., Wang, Z., Zhang, J., Zeng, Y., Wang, L.: Design and comparison of GaAs, GaAsP and InGaAlAs quantum-well active regions for 808-nm VCSELs. Opt. Express **19**, 1–13 (2011)

31. Kudrawiec, R.: Parametrization of energy gap at the Γ point and outside this point for dilute nitrides: $Ga_{1-y}In_yN_xP_{1-x}$ and $GaN_xAs_{1-x-z}P_z$ alloys. J. Appl. Phys. **105**, 063529 (2009)

32. Aissat, A., Nacer, S., Bensebti, M., Vilcot, J.P.: Investigation on the emission wavelength of GaInNAs/GaAs strained compressive quantum wells on GaAs substrates. Microelectron. J. **39**, 63–66 (2008)

33. Bousbih, F., Ben Bouzid, S., Chtourou, R., Charfi, F.F., Harmand, J.C., Ungaro, G.: Effect of nitrogen in the electronic structure of GaAsN and GaAsSb(N) compounds. Mater. Sci. Eng. C **21**, 251–254 (2002)

34. Zhao, C.Z., et al.: The parameters in the band-anticrossing model for $In_xGa_{1-x}N_yP_{1-y}$ before and after annealing. Sci. China Phys. Mech. Astron. **52**, 2160–2163 (2011)

Chapter 5
Organic Solar Cells: A Review

M. Benghanem and A. Almohammedi

Abstract In this chapter, we present different materials, devices structures, and different processing techniques for the fabrication of organic photovoltaic (OPV) cells. The manufacturer of these types of solar cells uses a new process to get the best efficiencies with low cost by using printing techniques and photoactive layers based on polymer materials. Also, many scientific research works are presented and some illustrations about processing techniques, such as roll-to-roll techniques, for the design of OPV cells are presented in this chapter.

Keywords Organic PV cells · Processing techniques · Materials · Polymer solar cells · Devices

5.1 Introduction

Organic photovoltaic (OPV) cells are considered as the third-generation solar cells which present new material such as organic polymer and tandem solar cells. In this work, we give a brief review of OPV cells with different classifications and applications. The structure of the device is described as well as the organic material in the active layer of the device. The fabrication of OPV cells at low cost is possible using new processing techniques, such as roll-to-roll technique under. The organic solar cells present a low efficiency and short lifetimes compared to inorganic solar cells. The organic solar cells present the advantage to be flexible, thin, lightweight, and versatility. The next section is an overview of OPV cells. Then, we present the working principle and device structures of organic solar cells. We describe the type of material used in the active layer and we focus on the roll-to-roll (R2R) processing of organic PV cells. Our contribution in the field of organic solar cells is to synthesize the nanocrystals of Pb-chalcogenides and study their opt electrical properties in

M. Benghanem (✉) · A. Almohammedi
Physics Department, Faculty of Science, Islamic University, Madinah, Saudi Arabia
e-mail: benghanem_mohamed@yahoo.fr

A. Mellit and M. Benghanem (eds.), *A Practical Guide for Advanced Methods in Solar Photovoltaic Systems*, Advanced Structured Materials 128,
https://doi.org/10.1007/978-3-030-43473-1_5

combinations with low band gap polymers such as PTB7and PTB7-Th, for applications to the hybrid solar cells. Nanocrystals of Pb-chalcogenides will be synthesized by wet chemical methods, by mixing of precursors of Pb and chalcogenides (S, Se, Te) together at high temperature (200–350 °C). Synthesis of nanocrystals would be confirmed by XRD analysis. HRTEM, SEM, will be employed to know the shape, size, and morphology of synthesized nanocrystals. For optical and electrical characterization, the thin films of synthesized material will be prepared in combination with low band gap conjugated polymers (PTB7 and PTB7-Th) by spin coater and contact electrodes deposited by thermal evaporation method. The obtained results would be analyzed by Mathcad software to extract the important parameters such as mobility, charge carrier density, traps, and activation energy. Finally, solar cells device will be fabricated in bulk heterojunction device configuration and the performance of the device will be analyzed by measuring the J-V characteristics and impedance spectra under standard conditions.

5.2   Organic Photovoltaic (PV) Cells

Photovoltaic devices convert solar radiation directly into electricity using solar cells such as silicon solar cells with efficiencies reach the value of 25% in research [1]. The second generation of thin-film solar cells using materials such as cadmium telluride (CdTe) and copper indium gallium selenide (CIGS) give an efficiencies around 19.6% for CIGS [1]. The third-generation PV cells use organic materials or polymers. The organic solar cells are characterized by low efficiencies and short lifetimes and present the advantage to be flexible, thin, and versatility.

There are different types of solar cells using many technologies which are dye-sensitized solar cells (DSSC), small-molecule organic solar cells, and organic solar cells based on polymers.

The efficiencies of polymer solar cells reached the value of 8.3% [1]. The organic PV cells are constituted by a bulk heterojunction of polymer and derivatives of carbon fullerene. The first scientific research on organic PV cells using small-molecule heterojunction has been done [2]. Later, the first dye/dye bulk heterojunction PV has been designed [3]. Also, the first polymer-C60 heterojunction PV was fabricated in a research laboratory [4]. Many other research works have been focused on the development of organic PV cells [5–16].

Organic PV cells using polymers present the advantage to be flexible, thin, and also there is solution processing based on coating and printing techniques [17, 18] such as slot-die coating [19], screen [20], gravure [21], and inkjet printing [22] on a flexible substrate. The technology of fabrication plays an important role to reduce the production cost such as the fabrication using vacuum-free R2R process on flexible substrates [23].

The principle inconvenient with this solution processing is the low efficiency of organic PV cells compared with inorganic PV cells such as silicon solar cells. In fact, the efficiency obtained in the laboratory is about 8.3% while the efficiency of the PV modules is around 3.5% [1]. Theoretically, the efficiency for single-junction cells reaches the value of 15% for tandem (multi-junction) solar cells [24, 25].

The lifetimes of organic PV cells are in excess of 1000 h of outdoor stability [26–29]. Actually, the research focuses on efficiency and stability for the same material constituting the organic solar cells.

5.3   Structures and Principles of Organic PV Cells

Organic solar cells are constituted by a bulk heterojunction structure which allows a better absorption of sunlight. The bulk heterojunction is a solution between electrons donor and acceptor. The donor is a polymer semiconductor like poly3-hexylthiophene (P3HT) and the acceptor can be polymers, semiconductor nanoparticles, metal oxides, or fullerenes such as PCBM in a photoactive layer [30].

In this section, we focus on fullerene material system based on the bulk hetero-junction concept [31]. Figure 5.1 shows the bulk heterojunction in organic solar cells.

An organic solvent is mixed with soluble donor and acceptor material, and then the solution is deposited on the substrate with conductive layers. A penetrated network is obtained by micro-phase separation after the evaporation of the solvent. After absorption of light, we get a large space between donor and acceptor which allow the separation of the charge.

The transport of these charge to anode and cathode is due to the continuity of paths to the electrodes. To increase the efficiency of organic PV cells, an adequate combination of material using adequate treatment processes is necessary during the phases of fabrication of OPV cells.

Figure 5.2 describes the principle of organic solar cells which is established in four steps for bulk heterojunction (BHJ) devices [12, 32]. The four steps are exciton generation, exciton diffusion, exciton dissociation, and charge carrier transport to the electrodes in a BHJ solar cell:

1. Exciton generation: This step absorbs photon and allows the electron to be in the lowest unoccupied molecular orbital (LUMO) and the hole moves to the highest occupied molecular orbital (HOMO). This is due to the Coulomb forces forming an exciton.

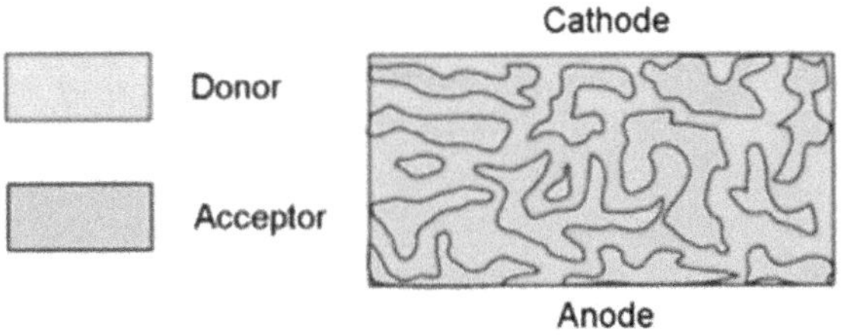

Fig. 5.1 Layer structures in organic solar cells

Fig. 5.2 Band diagram of
generated photocurrent

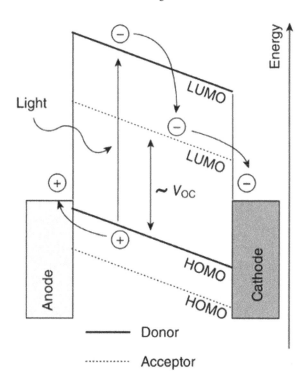

2. Exciton diffusion: The diffusion is done during the donor step to the interface of
   the donor and acceptor material. The concept of the bulk heterojunction of two
   mixed materials decreases the diffusion length and minimizes the decay rate of
   the exciton. The size of the two phases must be smaller than the length of the
   diffusion around 5–20 nm [33–36].
3. Exciton dissociation: The dissociation is done into a free electron and hole at the
   interface of donor and acceptor material.
4. Charge carrier transport: The charge carriers are transported via the donor and
   acceptor material. Negative charge are connected at the cathode, and the positive
   charge are collected at the anode. The generated photocurrent is due to applying a
   load to an external circuit. Organic materials characterized by a large absorption
   range can be synthesized and then influence the first step of the working principle.

Figure 5.3 presents the current density–voltage characteristics (J-V curve) for a solar
cell with different factors such as the open-circuit voltage (Voc in Volt), the short-
circuit current density (Jsc in mA/cm^2), the fill factor FF (%), and the maximum
power point MPP.

The efficiency of power conversion (η) is the first essential parameter and
expresses the ratio of the maximum output power ($P_M = I_{MPP} \cdot V_{MPP}$) generated
by the solar cell and the input incident power of the light (Pin) on a given active area
A:

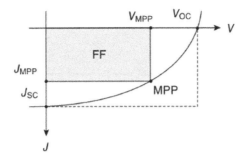

Fig. 5.3 Characteristic J-V curve of organic PV cell

$$n = \frac{I_{MPP} \cdot V_{MPP}}{P_M \cdot A} = FF \cdot \frac{I_{SC} \cdot V_{OC}}{P_M \cdot A} = FF.\frac{J_{SC} \cdot V_{OC}}{P_M}$$

where FF is defined as:

$$FF = \frac{I_{MPP} \cdot V_{MPP}}{I_{SC} \cdot V_{OC}} = \frac{J_{MPP} \cdot V_{MPP}}{J_{SC} \cdot V_{OC}}$$

The fill factor (FF) is a parameter which can determine the performance of the produced solar cells. In general, the acceptable organic solar cells correspond to the value of FF which should be greater or equal to 65%. Two other parameters characterize the performance of solar cells which are the series resistance R_S and the shunt resistance R_{Sh}. The series resistance represents the resistance at the interface in the layers, the conductivity of the semiconductors, and the electrodes. The shunt resistance corresponds to the defects in the layers and should be of high value.

The organic PV cells are constituted by a bulk heterojunction active layer with two electrodes as indicated in Fig. 5.4. One electrode is transparent, and generally, we use sputtering or evaporator on a transparent substrate (glass or polyethylene terephthalate) to get electrode with indium tin oxide (ITO). The last development of

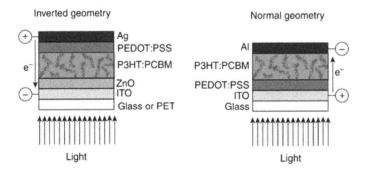

Fig. 5.4 Organic solar cells devices with a bulk heterojunction active layer in normal and inverted geometry

Table 5.1 The role of materials in normal and inverted geometries

Normal geometry

Substrate	Anode	Hole transport layer	Active layer	Electron transport layer	Cathode
Glass PET	**ITO**	**PEDOT: PSS** MoO_3 V_2O_3	**P3HT:PCBM**	TiO_x ZnO	LiF/**Al** LiF/Au

Inverted geometry

Substrate	Cathode	Electron transport layer	Active layer	Electron transport layer	Anode
Glass PET PEN	**ITO** **Ag-solid** **Al/Cr**	ZnO TiO_x CS_2-CO_3	**P3HT:PCBM**	**PEDOT: PSS**	**Ag** **Ag**-grid

PET = Polyethylene terephthalate; PEN = Polyethylene naphthalate; ITO = indium tin oxide; PEDOT:PSS = poly(3,4-ethylenediooxythiophene):poly(styrenesulfonate); MoO_3 = molybdenum trioxide; V_2O_5 = vanadium pentoxide; P3HT = poly(3-hexylthiophene); PCBM = [6]-phenyl C61 butyric acid methyl ester fullerene derivate); TiO_x = titanium oxide; nO = zinc oxide; LiF = lithium fluoride; Cs_2CO_3 = cesium carbonate; Al = aluminum; Au = gold; Ag = silver; and Cr = chromium

organic solar cells is to get devices with intermixed layers which are bulk hetero-junction of donor and acceptor. Actually, there are two references named normal and inverted geometry, necessary to build organic PV cells in the laboratory.

The transparent ITO electrode acts as the anode in the normal geometry, whereas in the inverted structure, it acts as a cathode. Table 5.1 shows the common materials in inverted and normal geometries.

The first device structure of organic solar cells has used the normal geometry [37, 38]. The use of vacuum for evaporating the cathode electrode on the top of the active layer is considered as inconvenient of this structure.

So, to avoid this vacuum steps, the researchers opted for inverted geometry by flipping the layer stack and adding a charge transport layer which gives the best solution to the process [17].

5.4 Materials

The active photoconversion layer and the hole transport layer PEDOT:PSS are in principle the only organic layers in an OSC. The active layer can be polymer-based, small-molecule-based, or a hybrid organic–inorganic structure. All other layers, except the substrate, are metals or metal oxides. Here, we briefly describe the several layer materials and focus on the organic polymer-based photoactive layer at the end.

5.4.1 Substrate and Front Electrode

Glass or polymeric materials such as polyethylene terephthalate (PET) or polyethylene naphthalate (PEN) are the basic substrates to build on the subsequent layer structure. PET or PEN is thin and flexible and makes it the first choice in large-scale R2R processing. Indium tin oxide (ITO) is widely used as a transparent electrode on glass or PET because of its excellent properties as a hole conductor. The drawback is the price and scarcity of indium. In addition, it uses vacuum-based processes for deposition, which shows up in a huge embodied energy of more than 80% in the final device [39]. Avoiding indium and finding alternative transparent conducting electrodes is highly demanding. One promising approach without using vacuum steps is the use of printed silver grids in combination with highly conductive poly(3,4-ethylenedioxythiophene) [40]. The chemical structure of PEDOT:PSS is shown in Fig. 5.5.

5.4.2 Intermediate Layers (ILs)

The intermediate layer (IL) between the active layer and the electrodes acts as a charge selective conductor, either blocking electrons or holes or conducting the opposite charge and vice versa. It might also improve the alignment of the energy levels and compensate for the roughness of the underlying surface to remove some of the shunts. A myriad of different materials have been studied, but it is beyond the scope of this chapter to mention them all [41]. In normal geometries, PEDOT:PSS is the most used electron-blocking material and is dissolved or dispersed in aqueous solution. It is spin-coated on ITO, forming a thin layer with thickness of 60–100 nm and is dried at approximately 150 °C for 5–10 min.

Most of the R2R produced OSCs are based on inverted structures, where electron-conducting materials are necessary as first IL. Typical materials are TiO_x or ZnO. They can easily be coated from solution-based nanoparticles or precursors of the metal oxides. An environmental side effect is obtained by using aqueous ZnO solution, and additionally, it alleviates the inflection point in the J-V curve caused by photodoping [42]. The layer thickness of the aqueous ZnO is in the range of 20 nm and needs heat treatment for 5–40 min at 140 °C to become insoluble and obtain electron-conducting characteristics.

The second IL between the active layer and back electrode acts as charge selecting layer similar to the first IL and has to have the opposite blocking characteristics than the first one. In OSCs with normal geometry, a second IL is not mandatory but materials like TiO_x and ZnO improve the efficiency and act as environmental barrier, inducing stability [43–45]. At the same time, the thin oxide layer can improve the optical absorption by shifting the field distribution inside the cell. Hole conducting PEDOT:PSS has been typically used in OSCs with inverted geometry, but MoO_3 and V_2O_5 have also been reported [46]. Applying water-based PEDOT:PSS on top

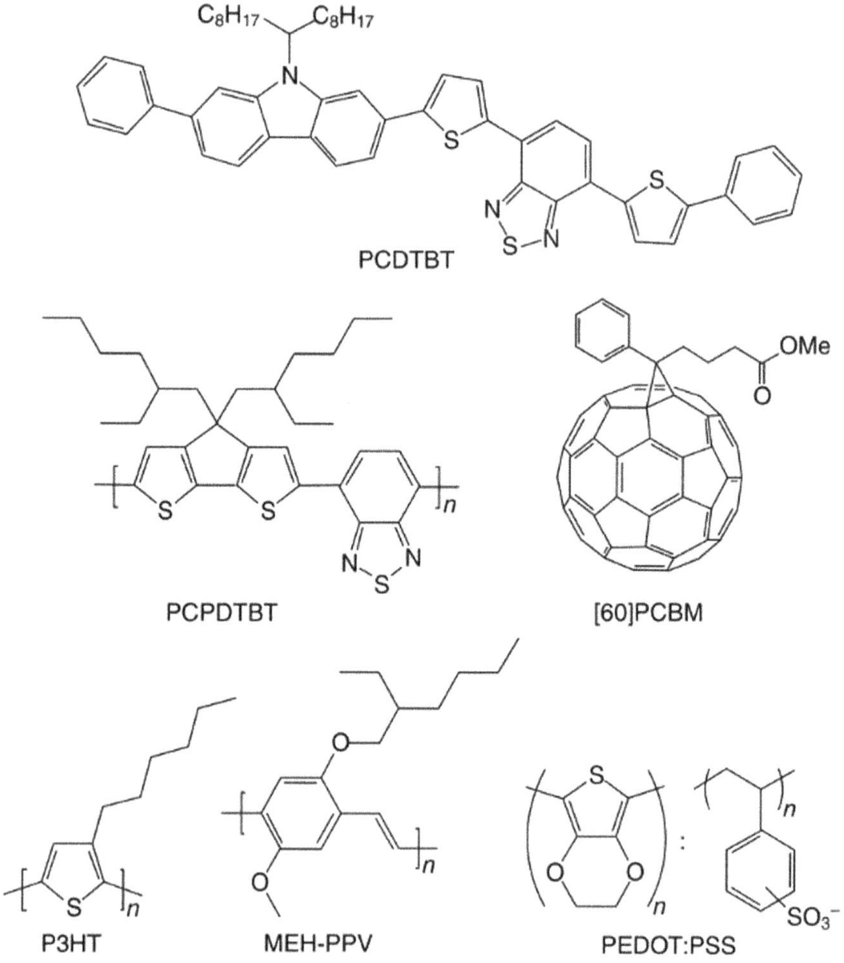

Fig. 5.5 Chemical structure of organic material used in OSC

of a hydrophobic PCBM:PCBM film, which causes dewetting and inhomogeneous layers, is almost impossible. Therefore, a special screen-printing formulation of PEDOT:PSS diluted in isopropanol is used for R2R coating [17]. For P3HT:PCBM-based OSCs, the interface between the active layer and PEDOT:PSS was found to be the weakest [47]. Delamination and thermomechanical stresses may result in poor device performance. Annealing time and temperature increase the adhesion in this interface.

5.4.3 Back Electrode and Encapsulation

A metallic back electrode completes the OSC structure and acts as either anode or cathode depending on the geometry. The most commonly used electrode material is aluminum, silver, and gold, but calcium has been reported too. Aluminum is often applied with a thin layer of lithium fluoride (2–10 nm), which improves contact to the active layer. Furthermore, it protects the active layer from damage during the evaporation [48]. In normal devices, the electrode is thermally evaporated and therefore not well suited for R2R processes. Silver electrodes can be easily applied by screen printing on inverted devices. The silver paste is commercially available and easy to process, but the influence of different solvents in the ink on the active layer has to be considered [49].

Encapsulation is necessary to prevent exposure to humidity and oxygen. A barrier material with a sufficient oxygen transmission rate (OTR) of at least $10-3$ cm^3 m^{-2} day^{-1} atm^{-1} and a water vapor transmission rate (WVTR) of at least $10-4$ gm^{-2} day^{-1} is required [50, 51]. Devices prepared on rigid substrate are typically encapsulated with glass or metal using epoxy. Flexible barrier foils are used for large-scale flexible encapsulation and can be applied by lamination with pressure-sensitive adhesives (PSA) [52]. Alternating layers of inorganic oxides such as SiOx and polymers in the barrier foil are used to achieve high OTR and WVTR.

5.4.4 Active Layer

Light absorption and charge carrier generation happen in the photoactive layer and therefore huge research efforts are being made to develop high-performance donor and acceptor materials. The main challenges are good stability, material abundance, cost efficiency, and large-scale processability, although not everything is fulfilled by one material at the moment. The following section outlines some of the more successful materials for OSCs extracted from countless reports. Table 5.2 summarizes the current state-of-the-art OSC laboratory-scale devices and shows the solar cell parameters for different materials and device structures.

The bulk heterojunction, interpenetrating network by blending donor and acceptor material, was introduced in 1995 and showed great improvements in charge separation and efficiencies [53]. They used 2-methoxy-5-(2-ethylhexyloxy)-polyphenylenevinylene (MEH-PPV) as electron donor and the soluble fullerene derivate [6]-phenyl C61 butyric acid methyl ester (PCBM) as electron acceptor in the intermixed active layer (Fig. 5.4). A bulk heterojunction active layer with material combinations of poly(3-hexylthiophene) (P3HT) (Fig. 5.4) and PCBM or PCBM is state of the art and well studied. Efficiencies are in the range of 5% [37, 38]. The mismatch of the absorption spectrum of P3HT with the solar emission spectrum

Table 5.2 Solar cell performance parameters of state-of-the-art bulk heterojunction OSC

Structure	V_{OC} (V)	J_{SC} (mA/cm^2)	FF (%)	PCE (%)
ITO/PEDOT: PSS/**P3HT: PCBM**/Al	0.63	9.5	68	5.0
ITO/PEDOT: PSS/**PCDTBT: PCBM**/BCP/Al	0.91	11.8	66	7.1
ITO/PEDOT: PSS/**PCDTBT: PCBM**/TiO$_x$/Al	0.88	10.6	66	6.1
ITO/PEDOT: PSS/**P3HT:bisPCBM**/Sm/Al	0.72	9.14	68	4.5
ITO/PEDOT: PSS/**PTB7: PCBM**/Ca/Al	0.74	14.5	69	7.4

Active layer is highlighted in bold

limits further improvement in efficiency. P3HT has a band gap of around 1.9 eV and absorbs only wavelength below 650 nm. The photon flux reaching the surface of the Earth has a maximum of approximately 1.8 eV (700 nm), and therefore, P3HT can harvest only 22.4% of available photons [54].

A way to overcome this physical barrier is the synthesis of polymer material with low band gaps collecting as many photons as possible. The offset of the HOMO and LUMO levels between donor and acceptor becomes important as well, whereas the open-circuit voltage of the device is defined by the difference between the energy level of the HOMO in the donor and the energy level of the LUMO in the acceptor. The lowest band gap of the two materials defines the maximum current. In the case of PCBM as acceptor, the optimum band gap has to be in the range of 1.2–1.7 eV. Absorption of more photons leads to potentially higher efficiencies.

The preparation of low band gap (LBG) polymers follows the donor–acceptor approach, in which the polymer backbone has electron-rich and electron-poor domains. One of the most promising and efficient LBG polymers is poly[2,6- (4,4-bis-(2-ethylhexyl)-4H-cyclopenta[2,1-b;3,4-b′]-dithiophene)-alt-4,7-(2,1,3-benzothiadiazole)] (PCPDTBT), which is based on a benzothiadiazole unit (acceptor) and a 4,4-bis (2-ethylhexyl)-4H-cyclopenta[2,1-b;3,4-b′] dithiophene unit (donor). That band gap is around 1.46 eV. Reported power conversion efficiencies are up to 4.5% in combination with PCBM and 6.5% with PCBM [55]. High efficiencies of up to 7.1% were achieved with the LBG polymerpoly[N-9″-hepta-decanyl-2,7-carbazole-alt-5,5-(4′,7′-di-2-thienyl-2′,1′,3′-benzothiadiazole)] (PCDTBT) and [12] PCBM dissolved in dichlorobenzene and 13% dimethyl sulfoxide (Chu et al. 2011). The cell had very good characteristics with VOC of 0.91 V, JSC of 11.8 mA/cm^2, and a FF of 66%. Further LBG polymers are reviewed in detail [54, 56].

The active layer is processed out of a blend solution of donor and acceptor. In case of P3HT:PCBM, the optimal ratio is around 1:1 [37] with concentration of 20–40 mg/ml. The range of solvents is large, but chlorobenzene or dichlorobenzene is typically used [57]. A certain dry layer thickness is achieved after deposition of the ink and evaporation of the solvent. The theoretical maximum JSC with a 5 μm thick active layer is calculated with 15.2 mA/cm^2 for an IQE value of 100%. For more realistic thicknesses of 400 nm and an IQE value of 80%, the JSC decreases to 10.2 mA/cm^2 [58].

Improving the efficiency of the device is done by thermal annealing the active layer. This can drastically change the structure and morphology of the material [59, 60]. P3HT can crystallize when the temperature is above the glass transition temperature. The concentration of PCBM has also an influence on the morphology upon annealing [61]. It can improve the phase separation due to PCBM cluster growth. Other possibilities to improve the morphology of the active layer are solvent vapor treatment [62] and additives in the active layer ink.

Polymers are made soluble by attaching solubilizing side chains such as alkyl groups onto the conjugated polymer backbone. They do not contribute to the light harvesting and make the material soft, which is related to the instability of OSCs [26]. After solution processing and drying, the side chains are useless and then can be removed. This interesting application can be achieved with thermocleavable materials. Thermocleavable ester groups are attached to the polymer backbone with a branched alkyl chain as solubilizing group. After heating to 200 °C, the solubilizing groups are eliminated leaving the polymer insoluble. Heterojunction devices with poly-3-(2-methyl-hexan-2-yl)-oxy-carbonylbithiophene (P3MHOCT) and C60 showed an improved stability after thermal treatment [26]. The P3MHOCT is converted to the more rigid and insoluble poly-3-carboxydithiophene (P3CT) at 200 °C, and with further heating at 300 °C, it is converted to native polythiophene (PT) (Fig. 5.6).

The cleaving is visible due to a color change from red over orange to purple-blue. OSC devices prepared with P3MHOCT:PCBM showed a large improvement in efficiency from P3MHOCT over P3CT to PT yielding at 1.5% for PT:PCBM [63]. Thermocleaving can be seen as a breakthrough in the processing of polymers and adopting this technique to other polymers offers several advantages such as only having the active components are in the final layer. Detailed information about thermocleavable polymers can be found in [13, 64, 65].

Polymer material development and new active layer concepts such as polymer–polymer solar cells, inorganic–organic hybrid solar cells, and nanostructured inorganics filled with polymer are extremely fast-moving fields of research. Tandem

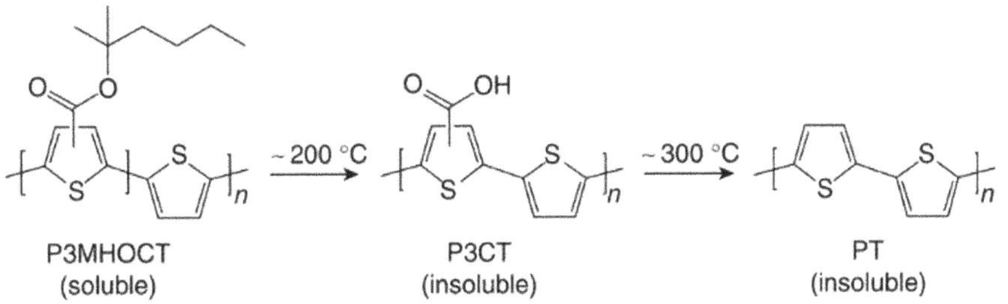

| P3MHOCT | P3CT | PT |
| (soluble) | (insoluble) | (insoluble) |

Fig. 5.6 Preparation of OT via a thermolytic reaction

OSCs improve the cell efficiency by stacking different band gaps materials on top of each other. The challenge is not only to find good material combinations to harvest as much photons as possible. From the processing point of view, the solvents and solubility of each subsequent material must match to prevent negative interaction like dissolving.

Thermocleaving and protecting the intermediate layer can improve the development of these multi-junction solar cells. Covering all of the further technologies in this chapter is not possible and we refer the reader to extensive review reports [12, 66, 67]. For the sake of completeness, it shall be mentioned that small-molecule organic solar cells [68, 69] are another group of organic solar cells and typically fabricated by vacuum processing in the preferred p-i-n structure [70, 71], employing either exciton blocking layers, or p-doped and n-doped electron transport layers. The intrinsic active layer is either a stacked planar heterojunction or a bulk heterojunction of donor and acceptor materials. The acceptor is typically the fullerene C60. Donor materials used are polyacenes [72], oligothiophenes [73], and metal phthalocyanines as benchmark materials [69]. Copper phthalocyanine (CuPc) and zinc phthalocyanine (ZnPc) are the most studied materials.

Red-absorbing fluorinated ZnPc (F4-ZnPc) can lead to 0.1–0.15 V higher open-circuit voltage than standard ZnPc. Combined with green-absorbing dicyanovinyl-capped sexithiophene (DCV6T) in a tandem cell structure, it can lead to an absorption over the whole spectrum with efficiencies up to 5.6% [74]. Efficiency of more than 8% in a tandem structure has already been achieved [1]. A drawback of small-molecule OSCs is the energy-intensive vacuum processing. Recent developments demonstrate solution-based small-molecule solar cells [75].

Efficiency of 6.7% was achieved with 5,5′-bis{(4-(7-hexylthiophen-2-yl) thiophen-2-yl)- [1, 2, 5] thiadiazolo [3,4-c]pyridine}-3,3′-di-2-ethylhexylsilylene-2,2′-bithiophene, DTS(PTTh2)2 as donor, and PCBM as acceptor dissolved in chlorobenzene. The normal structured bulk heterojunction device still comprises an evaporated Al cathode.

5.5   Roll-to-Roll (R2R) Processing of Organic PV Cells

The fabrication of organic solar cells is based on coating and printing techniques. The advantages of organic solar cells consist of the low-cost and very big production of such solar cells. Table 5.3 presents the performance of P3HT: PCBM solar cells, in which the active layer is fabricated with various process technologies such as roll-to roll (R2R) processing, slot-die [17, 19, 28], gravure [21, 76], screen printing [77], and inkjet [22, 78].

The preparation of standard solar cells in the laboratory using spin coater or evaporating systems presents the inconvenience which is the small active area and the loss of material during operation of fabrication by spin coater. The produced solar cells with such technology are limited due to the limitation of size, limited photocurrent, and low open-circuit voltage under 1 volts. To increase the productivity and to get

Table 5.3 Solar cell parameters of organic solar cells with P3HT: PCBM layers obtained by printing processes and coating

Process (active layer)	Active area (cm^2)	V_{OC} (V)	I_{SC} (mA)	FF (%)	PCE (%)
(a) Slot-die	4.80	3.62	6.86	44.00	2.33
(b) Slot-die	96	7.56	60.00	38	1.79
(c) Slot-die	13.2	6.48	>6	64	2.20
(d) Gravure	9.65	3.02	13.91	44.00	1.92
(e) Gravure	0.045	0.56	0.22	45.00	1.21
(f) Screen printing	1.44	0.59	21.07	29.78	2.59
(g) Inkjet	0.16	0.57	1.50	45.00	2.40
(h) Inkjct	0.09	0.628	0.9612	55.27	3.71

more output power of solar cells, we use R2R coating and printing techniques which allow the maximum use of material without loss. In general, in this technique, we use the inverter layer structure which allows to print the silver electrode with no vacuum.

The materials used for organic solar cells are soluble as printable ink and the used transparent conductive electrode ITO are evaporated on the flexible substrate. Due to the high cost of ITO and the embodied energy in organic solar cells is greater than 80% [39], many researches focus to avoid ITO and to use instead alternative materials like highly conductive polymers [79, 80]. For R2R processing, we use the flexible substrate and transparent foils such as polyethylene terephthalate (PET) and encapsulation using flexible barrier material allowing the final device lightweight and thin.

1. Full R2R, inverted structure, module of 8 serial connected cells [17].
2. Full R2R, inverted structure, module of 16 serial connected cells, average value over 600 modules, and max. PCE 2% [19].
3. Flatbed process, 11 serial connected cells, ITO-free, Cr/Al/Cr on flexible plastic substrate, and evaporated Au grid [28].
4. Normal geometry, flatbed process, and 5 serial connected cells [21].
5. Inverted structure, flatbed process, single cell, and evaporated Au electrode [76].
6. Normal structure, flatbed process, and evaporated Al electrode [77].
7. Normal structure and evaporated Al electrode [78].
8. Normal structure and evaporated LiF/Al electrode [22].

5.5.1 Serial Interconnected Device Structure

Organic solar cells module based on typical R2R consists of numerous serial interconnected single cells to get open-circuit voltages as illustrated in Fig. 5.7.

If we need 12 V in our PV system, we should connect 24 single cells with an open circuit of 0.5 V. The total photogenerated current is equal to the photogenerated

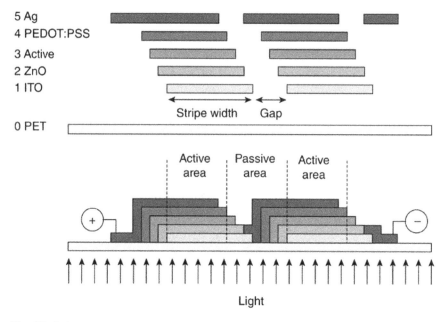

5 Ag
4 PEDOT:PSS
3 Active
2 ZnO
1 ITO

Stripe width   Gap

0 PET

Active area   Passive area   Active area

+   −

Light

Fig. 5.7 Serial interconnected of 2 single cells (inverted structure) (0) substrate PET, (1) ITO, (2) ZnO, (3) active layer, (4) PEDOT:PSS, and (5) Ag layer active area: generation of photocurrent

current of a single cell. This is due to the serial connection of identical solar cells which give the same short-circuit current. The production process using printing and coating allows homogeneous layer quality. During the industrial fabrication, we avoid any changes by fixing the process conditions in order to get similarity in the single cells. In research purposes, we can continuously change factors in R2R process. The conductivity of the electrode material depends on the width of the single cell, and upscaling reduces the photogenerated current.

The organic solar cells produced using R2R process have a stripe width of 4–20 mm for ITO (electrode material) and calculations of the size dependence [81]. Also, the small cell width decreases the relative active area of the module and some deposited material does not produce photocurrent. Many research [80, 82] focuses on monolithic design architecture with a high active to total area ratio by using highly conductive electrode material. The simplicity of manufacturing of this kind of device structure is an advantage since there is no required pattern.

5.5.2 Coating and Printing Processes

We describe in this section the process related to printing and coating for fabrication of OSCs devices. In fact, due to the facilities and availability of necessary materials and polymers for solution processing, it will be easy to transfer the laboratory-scale

Discrete

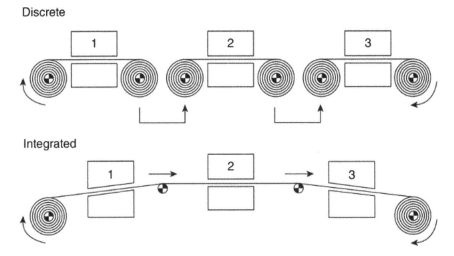

Integrated

Fig. 5.8 Comparison between integrated R2R and a discrete process

spin coating process to large scale such as R2R coating and printing processes. Additive deposition of material affects the price of the production of OSC devices. The technologies used in thin-film fabrication like photographic from industrial sectors can be suitable for the fabrication of OSCs.

The organic solar cell is based on different layers which may need various processing technologies, intermediate treatments, or different factors such as time of drying and web speed. To minimize the cost of steps in vacuum process, we prefer a structure using an inverted layer and printable electrodes. Also, it is desired to use an integrated R2R processing on a single factory line, but the different needs for each step allows discrete processing steps for the multi-layer OSC, as illustrated in Fig. 5.8. An optimization and best control represent an advantage of single-step processing. A total production can be stopped by any failure during the different steps of the process. An optimization can be done on different available machines of coating and printing following the needs of manufacturing. Printing methods are adequate to R2R processing, and different technologies can be used in a plant to fulfill the fabrication requirements for all layers, inks, and process parameters.

The principal need for the thin-film layer is homogeneity without pinholes over a large area [10, 83–86].

5.5.3 Slot-Die Coating

Slot-die coating provides smooth layers with homogeneous thicknesses in different directions. Figure 5.9 shows the system of coating. As it is described, the wet layer thickness is controlled by the coating speed and the flow rate of the supplied ink.

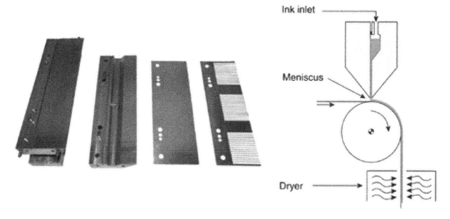

Fig. 5.9 Slot-die coating process (right) and the disassembled head including the stripe mask (left)

The coating head is precision engineered and it is not easy to get the same pressure distribution along the coating width. Multi-layer structures of organic solar cells are produced by displacing the head perpendicular to the web direction to align the stripes of materials in each process step. By inserting the stripes masks into the head, we get the stripes, forming a stripe-wide meniscus at the outlet of the ink. The ink viscosity adjusts the thickness of the mask and is in the range of 08–210 μm. Low-viscous polymer solutions for OSCs with viscosities below 20 MPa require masks with thicknesses of 25–55 μm. Due to the different wetting behaviors of the ink, the masks should be varied. The coating head lip is never in contact with the substrate. Pre-coated is not influenced by contact pressure such as in other printing technologies.

By using syringe pumps or piston, we pump the ink into the head. Higher viscous inks above 110 mPa may require pressure vessels with flow control equipment. The total coating process including the ink tank is a closed system which is useful for solvents. The wet layer thickness is fixed by the ink and web speed. The final dry layer thickness can be calculated with the following relation:

$$d = \frac{f}{Sw} \cdot \frac{c}{\rho}$$

d represents the thickness in cm, f represents the ink flow rate in cm³/min, c is the solid concentration in the ink in g/cm³, S is the web speed in cm/min, w is the coated width in cm, and ρ represents the density of the dried ink material in g/cm³.

Typical values of coating speeds for the fabrication of organic solar cell, with slot-die coating, are 0.4–2.5 m/min. This depends on the viscosity and the time of drying. The maximum drying temperature is fixed by substrate. Figure 5.10 presents a photon of slot-die coated organic solar cell modules with screen-printed silver electrodes [87]. Slot die can also be used for the coating of silver nanoparticle ink to

Fig. 5.10 Flexible OSCs modules fabricated by R2R slot-die coating with screen-printed silver electrodes

form electrode layers as a substitute for ITO [20]. Nanoparticle inks can be coated through slot-die coating technology. The coating factors and the properties of the ink should be optimized to reach the best quality of layer.

To study the material compositions in the processing parameter space [87], R2R slot-die coating is used as a research tool. Figure 5.11 shows the principle of differential pumped slot-die coating using two pumps for the material screening of ink compositions. This technology allows the material screening and characterization of OSCs with very small amounts of material in a R2R process. This late presents an advantage by comparison with individual spin coating. In fact, an optimization is obtained about the material ratios like P3HT to PCBM, and best layer thicknesses are much faster.

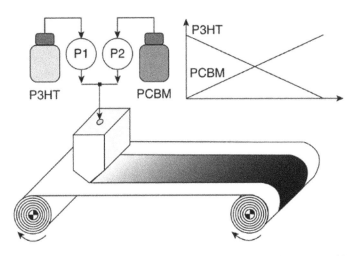

Fig. 5.11 Differential pumped slot-die coating for material screening of ink compositions

5.5.4 Screen Printing

For laboratory-scale research, we can use the technique of screen printing which is low-cost process. Also, this technique is used in the industrial process with rotary screen printing or continuous flatbed screen printing. The principle of this process is simple which is constituted by a screen and a squeegee. The screen has a cylindrical form and rotates around a fixed squeegee. The screen is a woven fabric made out of stainless steel or polyester covered with an emulsion layer. It is mounted to a metal frame under tension. Figure 5.12 illustrates the printing process where the screen is placed above the substrate distance by only a few millimeters. The ink is spread by moving the squeegee which moves over the screen with a sufficient pressure downward to the substrate. This allows the ink to left through the open areas of the screen behind on the substrate as the screen snaps back.

The obtained wet layer thickness is high compared to other printing techniques and the paste-like ink needs to be high viscous with thixotropic behavior. In order to avoid the clogging of the screen, the solvent should present a low volatility due to the exposition area of the ink to the environment. The principal factors of a screen are the mesh number which means mesh opening and the wire diameter. The theoretical paste volume (Vscreen) defines the wet layer thickness of the printed film. Vscreen represents the volume between the threads of the mask and the thickness of the emulsion. The volume of ink is influenced by process parameters, such as squeegee force, squeegee angle, snap-off distance, and ink rheology, summarized by the pickout ratio, k_p. The dry layer thickness d can be computed by the following relation:

$$d = V_{\text{screen}}.K_P.\frac{c}{\rho}$$

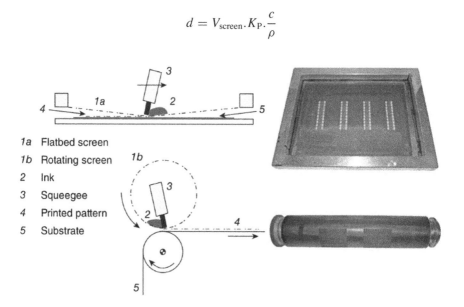

1a   Flatbed screen
1b   Rotating screen
2   Ink
3   Squeegee
4   Printed pattern
5   Substrate

Fig. 5.12 Principle of the flatbed and rotary screen printing process with corresponding screen

C represents the solid concentration in the ink in g/cm^3, and ρ is the density of the dried ink material in g/cm^3. Screen printing is used in the industry for graphical applications, but also for printing of conductors for flexible electronics. Also, it is used to print electrodes in the silicon solar cell industry. In the field of OSCs, screen printing of the active layer polymer MEH-PPV with an adequate rheology was successfully described [88]. Many scientific researches focused on screen printing for P3HT mixtures, which show the possibility of overcoming the challenges of ink rheology and large wet thickness [65, 77, 79].

Actually, screen printing allows the deposition of conductors like PEDOT:PSS and especially for silver ink in full solution processes [19, 80]. Also, thin lines are possible, which can be used to pattern grid structures with a honeycomb-like design, as used for monolithic OSCs [82]. The silver inks can be solvent, and the influence of different types of silver ink on the performance of OSCs has been studied [49]. Other, screen printing is used to pattern the ITO layer on PET by printing etch resist and further etching, stripping, and washing of the ITO layer [17].

5.5.5 *Gravure Printing*

Generally, in order to print graphical products with high speed (up to 20 m/s), we use the gravure printing. To get and homogenous layers with high-resolution patterning, we use the two-dimensional process and low-viscosity ink. Figure 5.13 shows the gravure printing process. The cells are filled in an ink bath or using a closed chambered doctor blade system. This is constituted by a chambered blade (doctor blade), a chromium-coated gravure cylinder, and a soft impression cylinder. To optimize the printing result, we can adjust the patterned gravure cylinder by controlling the volume (cm^3/m^2) of engraved cells, depth, width, density, and screen angle. The nip pressure allows the transfer of the ink to the substrate. The excess ink is left off just before the nip of the gravure and impression cylinder.

The pick-up volume and ink transfer rate from the cells to the substrate define the wet layer thickness. Different parameters affect the performance of the printing quality such as printing speed, pressure of the impression, and the ink rheology.

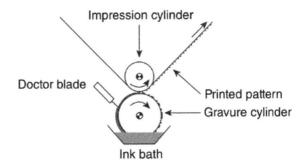

Fig. 5.13 Gravure printing process

The deposition of materials such as PEDOT:PSS and P3HT:PCBM in organic solar cells using normal structure is recommended with gravure printing process [21, 89].

Using high concentration of the ink (150 mg/ml in o-DCB), the speed of 18 m/min is obtained in this process. For a single cell, the obtained efficiency is 2.8%, and for the module (five connected stripes), the efficiency is 1.92%.

Also, the inverted structured solar cells with three gravure printed layers (TiO$_x$, P3HT:PCBM, and PEDOT:PSS) on patterned ITO-PET substrate using a speed of 40 m/min were also manufactured with success [76]. The obtained efficiency was 0.58% with a device area of 4.5 mm^2.

Also, R2R gravure printing process has been used with success for organic solar cells with P3HT:PCBM active layer [90]. The substrate was paper with a ZnO/Zn/ZnO layer as electrode and the active layer was gravure printed with a speed of 12 m/min. An efficiency of 1.31% (area: 9 mm^2 and an illumination of 600 W/m^2) has been obtained using PEDOT:PSS with R2R flexo printed process.

The fabrication of organic solar cells using first the process sheet-to-sheet to show the suitability of the R2R gravure process and electrodes was evaporated. The conversion to a full R2R process where all layers are gravure printed is still possible due to the high speed and the oven length required for drying and annealing.

5.5.6 Flexographic Printing

The flexographic printing is considered as R2R printing process with a speed up to 100 m/min by using a cylindrical plate carrying the printing pattern. The flexo system comprises a fountain roller which allows to fill the anilox roller with ink. The anilox roller allows a uniform thickness and equally to the printing plate cylinder. Excess ink is scraped off and it is transferred to the substrate running between the plate cylinder and impression cylinder, as shown in Fig. 5.14. A chambered doctor blade inking system can be used to avoid exposure of the ink to the atmosphere.

Fig. 5.14 Illustration of the flexographic printing process

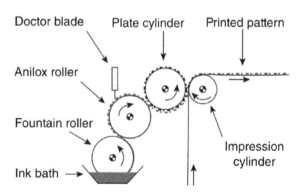

The flexographic printing has been used first for patterned PEDOT:PSS on top of P3HT:PCBM as indicated in the previous research [90].

The R2R printing of transformed PEDOT:PSS was realized with a speed of 30 m/min using an anilox cylinder (volume of 25 cm^3/m^2) with an efficiency of 1.31%. Flexotechnic was used for pre-wetting the surface of P3HT:PCBM with n-octanol prior to the coating of PEDOT:PSS [20]. The flexographic technic is applicable to organic photovoltaics with low volatility of the inks. Chambered doctor blade systems for the application of the ink are beneficial. It might be used for electrode production, either grids or full layer, because it can produce very thin layers. In another research work, silver grid structures (line width of 20–50 μm and distance of 0.8–2 mm) were successfully R2R printed at a speed of 5 m/min on PET [91].

5.5.7 Inkjet

Generally, inkjet is used for office applications, and actually, it is used for graphical applications such as R2R industrial process. One of the most applications in the industry is the ink droplets which are generated suitable on the need.

By heating up the ink and producing a tiny bubble to allow pushing out a droplet. The piezoelectric printhead is the most used in such applications. In order to generate a pressure to push out a drop, the use of the reverse piezoelectric effect is very useful as shown in Fig. 5.15.

The ideal drops are obtained by interaction of the ink and the inkjet head. In general, the ink requires a specific voltage to drive the piezoelectric printhead. The

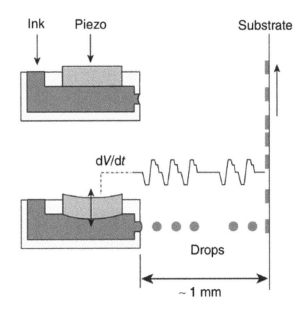

Fig. 5.15 Inkjet printing process using the reverse piezoelectric effect for generating pressure to push out a drop

volume of the generated drop depends on the type of print head. The inkjet printing process can reach high resolution over 1200 dpi. Inkjet ink must have a low evaporation rate in order to avoid drying at the nozzle. The viscosity is an important parameter and should be lower than 30 mPa. The inks are constituted by many solvents and surfactants and the inks can either be, solvent-based, aqueous or UV curable. Many other information used for inkjet applications have been described [92–94]. Many researches on organic solar cells use inkjet process [78, 95–97]. The achieved efficiency, using inkjet process, is about 3.7% [22].

5.6 Conclusion

Many companies begin working to fabricate organic solar cells which present the advantage to be flexible but low efficiency. Some applications use such type of solar cells and especially in building-integrated photovoltaics. Recent researches have focused to fabricate organic solar cells with different technic such as printing process. Some of the main research areas that have already been addressed are higher efficiencies, improvements in operational stability and lifetime, multi-junction and hybrid structures, low band gap polymers and controlled morphology. New organic materials have been used in organic solar cells to get the best efficiency and good performance. Different strategies in the development of new polymers for efficient organic solar cells have been presented in many research. Solar cells efficiencies around 8% that are published are manufactured with special conditions by testing very small cell sizes. The best challenge is to achieve high efficiencies on large areas under full R2R process. But hopefully, in the near future, the strong research efforts will make this happen, so that organic solar cells can compete against other solar cell technologies.

Acknowledgements We would like to thank the deanship of scientific research at Islamic University (Madinah, KSA) for supporting this first (Tamayouz) program of academic year 2018/2019, research project No.: 1/40. All collaboration works are gratefully acknowledged.

References

1. Green, M.A., et al.: Solar cell efficiency tables (Version 38). Prog. Photovoltaics Res. Appl. **19**(5), 565–572 (2011)
2. Tang, C.: Two-layer organic photovoltaic cell. Appl. Phys. Lett. **48**, 183 (1986)
3. Sariciftci, N.S., et al.: Observation of a photoinduced electron transfer from a conducting polymer (MEHPPV) onto C60. Synth. Met. **56**(2–3), 3125–3130 (1993)
4. Spanggaard, H., Krebs, F.C.: A brief history of the development of organic and polymeric photovoltaics. Sol. Energy Mater. Sol. Cells **83**(2–3), 125–146 (2004)
5. Brabec, C.J., Sariciftci, N.S., Hummelen, J.: Plastic solar cells. Adv. Func. Mater. **11**(1), 15–26 (2001)

6. Coakley, K., McGehee, M.D.: Conjugated polymer photovoltaic cells. Chem. Mater. **16**(23), 4533–4542 (2004)
7. Hoppe, H., Sariciftci, N.S.: Organic solar cells: an overview. J Mater Researd **19**(7), 1924–1945 (2004)
8. Dennler, G., Sariciftci, N.S.: Flexible conjugated polymer-based plastic solar cells: from basics to applications. Proc. IEEE **93**(8), 1429–1439 (2005)
9. Brabec, C.J., Dyakonov, V., Scherf, U.: Organic Photovoltaics, Wiley-VCH (2008)
10. Krebs, F.C.: Polymer Photovoltaics: A Practical Approach. SPIE Publications (2008)
11. Thompson, B., Frechet, J.M.J.: Polymer—fullerene composite solarcells. Angew. Chem. Int. Ed. **47**(1), 58–77 (2008)
12. Deibel, C., Dyakonov, V.: Polymer–fullerene bulk-heterojunction solar cells. Rep. Prog. Phys. **73**, 096401 (2010)
13. Helgesen, M., Søndergaard, R., Krebs, F.C.: Advanced materials and processes for polymer solar cell devices. J. Mater. Chem. **20**(1), 36–60 (2010)
14. Po, R., Maggini, M., Camaioni, N.: Polymer solar cells: recent approaches and achievements. J. Phys. Chem. C **114**(2), 695–706 (2010)
15. Servaites, J.D., Ratner, M.A., Marks, T.J.: Organic solar cells: a new look at traditional models. Energy Environ. Sci. **4**, 4410–4422 (2011)
16. Zhang, F., et al.: Recent development of the inverted configuration organic solar cells. Sol. Energy Mater. Sol. Cells **95**(7), 1785–1799 (2011)
17. Krebs, F.C.: Fabrication and processing of polymer solar cells: a review of printing and coating techniques. Sol. Energy Mater. Sol. Cells **93**(4), 394–412 (2009)
18. Søndergaard, R., et al.: Roll-to-roll fabrication of polymer solar cells. Mater. Today **15**(1–2), 36–49 (2012)
19. Krebs, F.C., Tromholt, T., Jørgensen, M.: Upscaling of polymer solar cell fabrication using full roll-to-roll processing. Nanoscale **2**(6), 873–886 (2010)
20. Krebs, F.C.: All solution roll-to-roll processed polymer solar cells free from indium-tin-oxide and vacuum coating steps. Org. Electron. **10**(5), 761–768 (2009)
21. Kopola, P., et al.: Gravure printed flexible organic photovoltaic modules. Sol. Energy Mater. Sol. Cells **95**(5), 1344–1347 (2011)
22. Eom, S.H., et al.: High efficiency polymer solar cells via sequential inkjet-printing of PEDOT:PSS and P3HT:PCBM inks with additives. Org. Electron. **11**(9), 1516–1522 (2010)
23. Nielsen, T., et al.: Business, market and intellectual property analysis of polymer solar cells. Sol. Energy Mater. Sol. Cells **94**(10), 1553–1571 (2010)
24. Koster, L., Mihailetchi, V., Blom, P.W.M.: Ultimate efficiency of polymer/fullerene bulk heterojunction solar cells. Appl. Phys. Lett. **88**, 093511 (2006)
25. Scharber, M.C., et al.: Design rules for donors in bulk-heterojunction solar cells—towards 10% energy-conversion efficiency. Adv. Mater. **18**(6), 789–794 (2006)
26. Krebs, F.C., Spanggaard, H.: Significant improvement of polymer solar cell stability. Chem. Mater. **17**(21), 5235–5237 (2005)
27. Hauch, J., et al.: Flexible organic P3HT: PCBM bulk-heterojunction modules with more than 1 year outdoor lifetime. Sol. Energy Mater. Sol. Cells **92**(7), 727–731 (2008)
28. Zimmermann, B., Würfel, U., Niggemann, M.: Longterm stability of efficient inverted P3HT: PCBM solar cells. Sol. Energy Mater. Sol. Cells **93**(4), 491–496 (2009)
29. Voroshazi, E., et al.: Long-term operational lifetime and degradation analysis of P3HT:PCBM photovoltaic cells. Sol. Energy Mater. Sol. Cells **95**(5), 1303–1307 (2011)
30. Basics of OSC: Organic solar cells, [Online]. Available: https://www.iapp.de/iapp/agruppen/osol/?Research:Organic_Solar_Cells:Basics_of_OSC
31. Dam, H.F., Larsen-Olsen, T.T.: How do polymer solar cells work [Online]. Available: http://plasticphotovoltaics.org/lc/lc-polymersolarcells/lc-how.html
32. Blom, P.W.M., et al.: Device physics of polymer: fullerene bulk heterojunction solar cells. Adv. Mater. **19**(12), 1551–1566 (2007)
33. Halls, J., et al.: Exciton diffusion and dissociation in a poly (p-phenylenevi-nylene)/C60 heterojunction photovoltaic cell. Appl. Phys. Lett. **68**, 3120 (1996)

34. Haugeneder, A., et al.: Exciton diffusion and dissociation in conjugated polymer/fullerene blends and heterostructures. Phys. Rev. B **59**(23), 15346 (1999)
35. Pettersson, L.A.A., Roman, L.S., Inganäs, O.: Modeling photocurrent action spectra of photovoltaic devices based on organic thin films. J. Appl. Phys. **86**, 487 (1999)
36. Piris, J., et al.: Photogeneration and ultrafast dynamics of excitons and charges in P3HT/PCBM blends. J. Phys. Chem. C **113**(32), 14500–14506 (2009)
37. Li, G., et al.: High-efficiency solution processable polymer photovoltaic cells by self-organization of polymer blends. Nat. Mater. **4**, 864–868 (2005)
38. Ma, W., et al.: Thermally stable, efficient polymer solar cells with nanoscale control of the interpenetrating network morphology. Adv. Func. Mater. **15**(10), 1617–1622 (2005)
39. Espinosa, N., et al.: A life cycle analysis of polymer solar cell modules prepared using roll-to-roll methods under ambient conditions. Sol. Energy Mater. Sol. Cells **95**(5), 1293–1302 (2011)
40. Galagan, Y., Rubingh, J.M., et al.: ITO-free flexible organic solar cells with printed current collecting grids. Solar Energy Mater. Solar Cells, **95**(5), 1339–1343 (2011a)
41. Po, R., et al.: The role of buffer layers in polymer solar cells. Energy Environ. Sci. **4**(2), 285–310 (2011)
42. Søndergaard, R., et al.: Fabrication of polymer solar cells using aqueous processing for all layers including the metal back electrode. Adv. Energy Mater. **1**(1), 68–71 (2010)
43. Gilot, J., et al.: The use of ZnO as optical spacer in polymer solar cells: theoretical and experimental study. Appl. Phys. Lett. **91**(11), 113520–4 (2007)
44. Hayakawa, A., et al.: High performance polythiophene/fullerene bulk-hetero-junction solar cell with a TiO$_x$ hole blocking layer. Appl. Phys. Lett. **90**(16), 163517–4 (2007)
45. Lee, K., et al.: Air-stable polymer electronic devices. Adv. Mater. **19**(18), 2445–2449 (2007)
46. Huang, J.-S., Chou, C.-Y., Lin, C.-F.: Efficient and air-stable polymer photo-voltaic devices with WO$_3$-V$_2$O$_5$ mixed oxides as anodic modification. Electron Device Lett. IEEE **31**(4), 332–334 (2010)
47. Dupont, S.R., et al.: Interlayer adhesion in roll-to-roll processed flexible inverted polymer solar cells. Sol. Energy Mater. Sol. Cells **97**, 171–175 (2012)
48. Brabec, C.J., et al.: Effect of LiF/metal electrodes on the performance of plastic solar cells. Appl. Phys. Lett. **80**(7), 1288 (2002)
49. Krebs, F.C., Søndergaard, R., Jørgensen, M.: Printed metal back electrodes for R2R fabricated polymer solar cells studied using the LBIC technique. Sol. Energy Mater. Sol. Cells **95**(5), 1348–1353 (2011)
50. Dennler, G., Lungenschmied, C., Neugebauer, H.: Flexible, conjugated polymer-fullerene-based bulk-heterojunction solar cells: Basics, encapsulation and integration. J. Mater. Res. **20**(12), 3224–3233 (2005)
51. Lungenschmied, C., Dennler, G., Neugebauer, H.: Flexible, long-lived, large-area, organic solar cells. Sol. Energy Mater. Sol. Cells **91**(5), 379–384 (2007)
52. Tanenbaum, D.M., et al.: Edge sealing for low cost stability enhancement of roll-to-roll processed flexible polymer solar cell modules. Sol. Energy Mater. Sol. Cells **97**, 157–163 (2012)
53. Yu, G., et al.: Polymer photovoltaic cells: enhanced efficiencies via a network of internal donor-acceptor heterojunctions. Science **270**(5243), 1789–1791 (1995)
54. Bundgaard, E., Krebs, F.C.: Low band gap polymers for organic photovoltaics. Sol. Energy Mater. Sol. Cells **91**(11), 954–985 (2007)
55. Boland, P., Lee, K., Namkoong, G.: Device optimization in PCPDTBT:PCBM plastic solar cells. Sol. Energy Mater. Sol. Cells **94**(5), 915–920 (2010)
56. Kumar, P., Chand, S.: Recent progress and future aspects of organic solar cells. Prog. Photovoltaics Res. Appl. **20**, 377–415 (2011)
57. Dang, M.T., et al.: Polymeric solar cells based on P3HT:PCBM role of the casting solvent. Sol. Energy Mater. Sol. Cells **95**(12), 3408–3418 (2011)
58. Dennler, G., Scharber, M.C., Brabec, C.J.: Polymer-fullerene bulk-hetero-junction solar cells. Adv. Mater. **21**(13), 1323–1338 (2009)

59. Padinger, F., Rittberger, R., Sariciftci, N.S.: Effects of postproduction treatment on plastic solar cells. Adv. Func. Mater. **13**(1), 85–88 (2003)

60. Benanti, T., Venkataraman, D.: Organic solar cells: an overview focusing on active layer morphology. Photosynth. Res. **87**(1), 73–81 (2006)

61. Chirvase, D., Parisi, J., Hummelen, J.: Influence of nanomorphology on the photovoltaic action of polymer–fullerene composites. Nanotechnology **15**, 1317–1323 (2004)

62. Zhao, Y., et al.: Solvent-vapor treatment induced performance enhancement of poly (3-hexylthiophene): methanofullerene bulk-heterojunction photovoltaic cells. Appl. Phys. Lett. **90**, 043504 (2007)

63. Gevorgyan, S.A., Krebs, F.C.: Bulk heterojunctions based on nativepolythiophene. Chem. Mater. **20**(13), 4386–4390 (2008)

64. Petersen, M., Gevorgyan, S.A., Krebs, F.C.: Thermocleavable low band gap polymers and solar cells therefrom with remarkable stability toward oxygen. Macromolecules **41**(23), 8986–8994 (2008)

65. Jørgensen, M., Hagemann, O., Alstrup, J.: Thermo-cleavable solvents for printing conjugated polymers: application in polymer solar cells. Sol. Energy Mater. Sol. Cells **93**(4), 413–421 (2009)

66. Sista, S., et al.: Tandem polymer photovoltaic cells—current status, challenges and future outlook. Energy Environ. Sci. **4**(5), 1606 (2011)

67. Weickert, J., et al.: Nanostructured organic and hybrid solar cells. Adv. Mater. **23**(16), 1810–1828 (2011)

68. Peumans, P., Yakimov, A., Forrest, S.R.: Small molecular weight organic thin-film photodetectors and solar cells. J. Appl. Phys. **93**(7), 3693–3723 (2003)

69. Rand, B.P., et al.: Solar cells utilizing small molecular weight organic semiconductors. Prog. Photovoltaics Res. Appl. **15**(8), 659–676 (2007)

70. Maennig, B., et al.: Organic p-i-n solar cells. Appl. Phys. A Mater. Sci. Process. **79**(1), 1–14 (2004)

71. Riede, M., et al.: Small-molecule solar cells—status and perspectives. Nanotechnology, **19**(42), 424001 (2008)

72. Yoo, S., Domercq, B., Kippelen, B.: Efficient thin-film organic solar cells based on pentacene/C60 heterojunctions. Appl. Phys. Lett. **85**(22), 5427 (2004)

73. Schulze, K., et al.: Efficient vacuum-deposited organic solar cells based on a new low-bandgap oligothiophene and fullerene C60. Adv. Mater. **18**(21), 2872–2875 (2006)

74. Riede, M., et al.: Efficient organic tandem solar cells based on small molecules. Adv. Func. Mater. **21**(16), 3019–3028 (2011)

75. Zhang, L., et al.: Triisopropylsilylethynyl-functionalized dibenzo[def, mno]chrysene: a solution-processed small molecule for bulk heterojunction solar cells. J. Mater. Chem. **22**(10), 4266 (2012)

76. Voigt, M.M., et al.: Gravure printing for three subsequent solar cell layers of inverted structures on flexible substrates. Sol. Energy Mater. Sol. Cells **95**, 731–734 (2011)

77. Zhang, B., Chae, H., Cho, S.: Screen-printed polymer: fullerene bulk-heterojunction solar cells. Jap. J. Appl. Phys. **48**(2), 020208-1–020208-3 (2009)

78. Lange, A., et al.: A new approach to the solvent system for inkjet-printed P3HT:PCBM solar cells and its use in devices with printed passive and active layers. Sol. Energy Mater. Sol. Cells **94**(10), 1816–1821 (2010)

79. Krebs, F.C., Jørgensen, M., Norrman, K., Hagemann, O., et al.: A complete process for production of flexible large area polymer solar cells entirely using screen printing—first public demonstration. Sol. Energy Mater. Sol. Cells **93**(4), 422–441 (2009)

80. Galagan, Y., Rubingh, J.-E., et al.: ITO-free flexible organic solar cells with printed current collecting grids. Solar Energy Mater. Solar Cells, **95**(5), 1339–1343 (2011a)

81. Hoppe, H., Seeland, M., Muhsin, B.: Optimal geometric design of monolithic thin-film solar modules: architecture of polymer solar cells. Sol. Energy Mater. Sol. Cells **97**, 119–126 (2012)

82. Manceau, M., et al.: ITO-free flexible polymer solar cells: from small model devices to roll-to-roll processed large modules. Org. Electron. **12**(4), 566–574 (2011)

83. Gutoff, E.B., Cohen, E.D.: Coating and drying defects, 2nd edn. Wiley, Hoboken, NJ (2006)
84. Tracton, A.A.: Coatings Technology. CRC Press, Boca Raton, FL (2007)
85. Kipphan, H.: Handbook of print media. Springer Verlag, Berlin, Heidelberg, New York (2001)
86. Krebs, F.C., Gevorgyan, S.A., Alstrup, J.: A roll-to-roll process to flexible polymer solar cells: model studies, manufacture and operational stability studies. J. Mater. Chem. **19**(30), 5442–5451 (2009)
87. Alstrup, J., et al.: Ultra fast and parsimonious materials screening for polymer solar cells using differentially pumped slot-die coating. ACS Appl. Mater. Interfaces. **10**(2), 2819–2827 (2010)
88. Krebs, F.C., et al.: Large area plastic solar cell modules. Mater. Sci. Eng., B **138**(2), 106–111 (2007)
89. Kopola, P., et al.: High efficient plastic solar cells fabricated with a high-throughput gravure printing method. Sol. Energy Mater. Sol. Cells **94**(10), 1673–1680 (2010)
90. Hübler, A.C., et al.: Printed paper photovoltaic cells. Adv Energy Mater **1**(6), 1018–1022 (2011)
91. Deganello, D., et al.: Patterning of micro-scale conductive networks using reel-to-reel flexographic printing. Thin Solid Films **518**(21), 6113–6116 (2010)
92. Derby, B.: Inkjet printing of functional and structural materials: fluid property requirements, feature stability, and resolution. Annu. Rev. Mater. Res. **40**, 395–414 (2010)
93. Pond, S.F.: Inkjet technology and product development strategies. Torrey Pines Res, Carlsbad (2000)
94. Magdassi, S.: The Chemistry of Inkjet Inks. World Scientific Publishing, Singapore (2010)
95. Hoth, C.N., et al.: High photovoltaic performance of inkjet printedpolymer: fullerene blends. Adv. Mater. **19**(22), 3973–3978 (2007)
96. Hoth, C.N., et al.: Printing highly efficient organic solar cells. Nano Lett. **8**(9), 2806–2813 (2008)
97. Kim, J.-M., et al.: Polymer based organic solar cells using ink-jet printed active layers. Appl. Phys. Lett. **92**, 033306 (2008)

Chapter 6
Doped ZnO Thin Films Properties/Spray Pyrolysis Technique

F. Z. Bedia, A. Bedia, N. Maloufi, and M. Aillerie

Abstract In this contribution, the effect of different dopants (Al, Sn, and Cu) on the structure, texture and optical properties of ZnO thin films was investigated. Al-doped ZnO (AZO), Sn-doped ZnO (TZO) and Cu-doped ZnO (CZO) thin films are synthesized by chemical spray pyrolysis technique on glass substrates. The so-obtained films crystallized in hexagonal wurtzite polycrystalline structure. The pole figures show that all the thin films have (0002) as the preferred orientation along the c-axis with the highest level was obtained in TZO thin film. The morphology film was significantly affected by the doping type. The transmittance spectra of all the films point out highly transparent in the visible range with an average transmittance higher than 80% for TZO and AZO films but with an average transmittance equal to about 70% for CZO film. Furthermore, the optical bandgap values were determined by the Tauc's law and were found to be 3.30 eV, 3.28 eV and 3.27 eV for AZO, TZO, and CZO thin films, respectively. The Urbach energy of the films was also calculated.

Keywords Doped ZnO · Thin films · Spray pyrolysis · Pole figures · Optical properties

F. Z. Bedia · A. Bedia (✉)
URMER, Abou-Bakr Belkaid University, Tlemcen 13000, Algeria
e-mail: aphy_bedia@yahoo.fr

N. Maloufi
LEM3, UMR CNRS 7239, Université de Lorraine, Metz 57045, France

DAMAS, Université de Lorraine, Metz 57045, France

M. Aillerie
LMOPS-EA 4423, Université de Lorraine, Metz 57070, France

LMOPS, Supelec, Metz 57070, France

© The Editor(s) (if applicable) and The Author(s), under exclusive license   107
to Springer Nature Switzerland AG 2020
A. Mellit and M. Benghanem (eds.), *A Practical Guide for Advanced Methods in Solar Photovoltaic Systems*, Advanced Structured Materials 128,
https://doi.org/10.1007/978-3-030-43473-1_6

6.1 Introduction

Zinc oxide (ZnO) has potential applications for a variety of optoelectronic devices including photodetectors [1], light-emitting diodes (LEDs) and laser devices [2]. By else, ZnO successfully replaces tin and indium oxides as transparent conductive films and window materials in solar cell applications due to its high optical transmittance in the visible light region and due to its high stability [3]. Moreover, ZnO films are extensively used as gas sensors because of their good sensing property and cost-effective route of synthesis [4]. Recently, the most investigated application of zinc oxide is in bio-sensing because of its high isoelectric point, biocompatibility and fast electron transfer kinetics [5]. However, ZnO is a II–VI compound semiconductor, and it crystallizes in wurtzite, zinc blend and rock salt structures. At ambient conditions, the stable phase is the hexagonal wurtzite crystal structure (a $\approx$ 3.249 Å; c $\approx$ 5.205 Å) [6]. This hexagonal structure is composed of alternating planes of tetrahedrally coordinated O^{2-} (1/3; 2/3; 0.38) and Zn^{2+} (1/3; 2/3; 0) ions along the c-axis. Compared to other II–VI semiconductors such as ZnSe and ZnS or to III–V compounds such as GaN, ZnO combines many advantages: (1) ZnO is a n-type semiconductor and its principal donor centers are usually identified as oxygen vacancies, zinc interstitials or complex defects [7, 8]; (2) ZnO possesses a direct wide bandgap semiconductor (3.37 eV) with a very large exciton binding energy of 60 meV at room temperature (compared to GaN: 25 meV, ZnSe: 22 meV, ZnS: 40 meV) [9, 10]; (3) ZnO occurs in many forms such as bulk single crystal, powder, thin film, nanowires and nanotubes. ZnO thin films were considered as an interesting alternative to replace indium tin oxide (ITO) in TCO layers owing to the low cost, natural availability and non-toxicity of this oxide [11, 12]. Furthermore, ZnO thin films are widely studied because they can be grown on different substrates at low temperature. These films are optically transparent in the visible spectrum. Depending on the specific process used for their growth, ZnO films can have widely varying electrical resistivity from 10^{-4} to 10^{12} Ω cm [13, 14]. Thereby, the elaboration of ZnO films is of intense interest, and diverse processes such as sol–gel, wet chemical method, physical and chemical vapor deposition, vapor–liquid–solid method and sputtering are presented in the literature to grow large varieties of ZnO thin films [15–17]. Compared to other techniques, the spray pyrolysis method [18], which was initially developed for conductive oxide deposition on solar cells applications, has several advantages: the spray pyrolysis method allows large area deposition; it is flexible for process modifications; it is simple and non-expensive. Furthermore, to optimize the physical properties requested by the large panel of applications, the ZnO thin films produced with this technique can be doped with a great variety of elements. ZnO thin films are often deliberately doped with group III elements (B, Al, In, Ga) and/or group IV elements (Pb, Sn). In addition, transition elements such as Cu^{2+}, Fe^{3+}, Co^{3+}, Mn^{2+} are also used to change the ZnO optical and electrical properties [19]. Paraguay et al. [20] studied the relationships between microstructure and a variety of dopants (Al, Cu, Fe, In and Sn) in ZnO films grown by spray pyrolysis. Further studies by Han et al. [21] examined the functional properties of undoped, and Al-, Fe-, Ni- and Sn-doped ZnO films are prepared by co-precipitation method. The influence of the concentration of dopants, such as Ni, Mo, V, F, Cu, Sn on the structure and/or the optical constants of ZnO thin films, has been studied by various research groups [22–27]. In this chapter, Al-, Sn- and Cu-doped ZnO thin films were grown on glass substrates by the spray pyrolysis method. The microstructure of these films was characterized by high-resolution scanning electron microscopy (FESEM). Measurements of transmittance in the UV–VIS–NIR range were used to calculate the optical parameters.

6.2 Experimental Process

6.2.1 Spray Pyrolysis-Based Synthesis

By spray pyrolysis process in ambient atmosphere, we have synthesized doped ZnO thin films. The spraying setup used was a home-made system that we have developed to perform a good spraying sample. Doped ZnO thin films have been deposited onto the glass substrates at 350 °C. The started solution was composed of 0.08 M of dehydrated zinc acetate (Zn $(CH_3COO)_2 \cdot 2H_2O$) diluted in methanol (CH_3OH). The aluminum chloride ($AlCl_3$), tin chloride dehydrated ($SnCl_2$, $2H_2O$) and copper chloride dehydrated ($CuCl_2, 2H_2O$) were added to the starting solution as a dopant. The Al/Zn, Cu/Zn and Sn/Zn atomic ratio, estimated in the staring solution, were used 1 wt.% for Al, Cu and Sn. It is to be noted that special care was taken for the cleaning of the glass substrate. Glass plates were immersed in ultrasonic successive bathes of various solutions of ethanol and acetone during 20–30 min and finally washed by distilled water, in order to achieve the final cleaning. After this growing phase, a post-growth treatment was carried out in order to properly finalize the films growth procedure by eliminating residual organic products. This post-growth phase consists to anneal the samples at 350 °C for one hour. The thickness of the so-obtained films was determined by the Swanepoel's envelope method [28] applied to the transmittance data, and the average thickness of all the films is equal to 300 nm.

6.2.2 Measurements

Following the previous phase linked to the growing process, we have first character-ized at room temperature, the quality and the structure of the so-obtained doped ZnO thin films by X-ray diffraction (XRD). The XRD technique was based on a four-circle Brucker D8 advanced diffractometer equipped with a rotating anode using Cu Kα radiation ($\lambda = 1.54059$ Å). The detector is a curved detector (INEL, CPS120), which allows to measure many planes on the 120° (from 5° to 125°) in unique acquisition time. The instrumental broadening in 2θ geometry is 0.004°. The pole figures were plotted to evaluate the texture of the film and provide information on the preferential orientation distribution in the material. The measurements were carried out with a fixed incident beam angle ($\omega = 23.8°$), which correspond to the Bragg position of the $(10\bar{1}2)$ planes. The pole figures were obtained by fitting the diffraction spectra at each Φ/Ψ position for the hexagonal $(10\bar{1}0)$, (0002), $(10\bar{1}1)$, $(10\bar{1}2)$, $(11\bar{2}0)$ and $(10\bar{1}3)$ planes. The MTEX [29] software was used to calculate the orientation function of the material. The morphology of the synthesized doped ZnO films was characterized in a high-resolution thermal field emission gun scanning electron microscope "FESEM" Zeiss Supra 40. The transmittance (T) measurement of the films were studied according to the UV–Visible–NIR spectrum recorded at room temperature.

6.3 Results and Discussion

6.3.1 Morphological Properties

The morphology of doped ZnO thin film was characterized by analysis of the SEM images as shown in Fig. 6.1. SEM images of AZO, TZO and CZO thin films show that all the films are dense and nano-crystallized. For all the samples, the distribution of the grain sizes and shapes appear uniform. Nanoparticles highly aggregated are observed in TZO thin film with hexagonal-shaped grains, and lateral size is in the range of ~ 100–150 nm, as seen in Fig. 6.1a. The morphology AZO thin film changes into a pencil shaped with a high density of small grains, as shown in Fig. 6.1b. However, the morphology of CZO thin film shows the appearance of granular structures composed of small grains that appear somehow aggregated with porosity around (Fig. 6.1c).

6.3.2 Structural Properties

The crystal structure and orientation of doped ZnO thin films were investigated by XRD with experimental parameters reported in the previous section. The diffractogram of Fig. 6.2 shows sharp and well-resolved peaks, which are a clear signature of the good crystallinity of the films. The diffractograms associated with JCPDS 36-1451 card [30] show that all the films are polycrystalline in hexagonal wurtzite structure. No additional diffraction peaks related to other impurity phases are observed. These results indicate that the substitution of Zn ions by Sn, Al or Cu ions does not change the structure of ZnO film. By else, diffractograms analysis show that all the films have multiple orientations such as $(10\bar{1}0)$, (0002), $(10\bar{1}1)$, $(10\bar{1}2)$, $(11\bar{2}0)$ and $(10\bar{1}3)$.

The structure of the doped ZnO thin films can be observed through lattice constants calculations according to the Bragg formula $\lambda = 2d_{hkl} \sin \alpha$. The lattice parameters

Fig. 6.1 High-resolution
SEM micrographs of doped
ZnO thin films: **a** AZO,
b TZO and **c** CZO

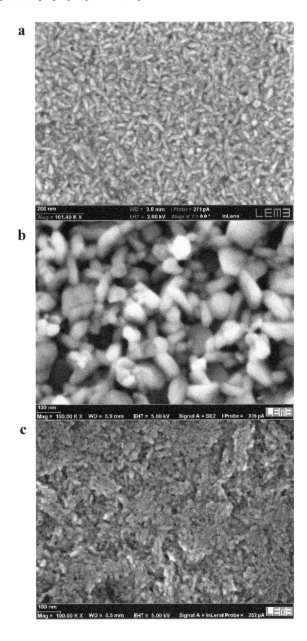

a and c corresponding to the lattice constants can be calculated via the following
formula [31]:

$$\frac{1}{d_{hkl}^2} = \frac{4}{3}\left[\frac{h^2 + hk + k^2}{a^2}\right] + \frac{l^2}{c^2} \tag{6.1}$$

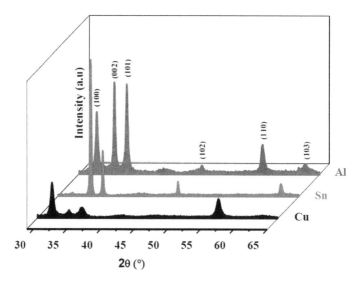

Fig. 6.2 XRD pattern of the Al, Sn and Cu doped ZnO thin films

Table 6.1 Calculated values of lattice parameters of AZO, TZO and CZO thin films

Thin films	a (Å)	c (Å)	a/c	u	l	ε_{zz}
AZO	3.245	5.201	0.623	0.379	1.975	−0.104
TZO	3.243	5.193	0.624	0.380	1.973	−0.262
CZO	3.235	5.213	0.620	0.378	1.972	+0.123

The deduced lattice parameters a and c and the a/c ratio for all the films are summarized in Table 6.1.

Bond length (l) and positional parameters (u) are calculated using the following formula [32]

$$l = \sqrt{\frac{a^2}{3} + \left(\frac{1}{2} - u\right)^2 \times c^2} \qquad (6.2)$$

$$u = \frac{a^2}{3c^2} + 0.25 \qquad (6.3)$$

The lattice parameters deduced change slightly with doping. This could be due to the stress formation induced by the ionic radii difference between Zn^{2+} (0.74 Å), Sn^{4+} (0.69 Å), Al^{3+} (0.53 Å) and Cu^{2+} (0.75 Å) [18, 33–35].

In thin film grown on glass substrate, the strain originates likely from intrinsic rather than thermal origins [36]. This strain can be positive (tensile) or negative (compressive) [37]. For ZnO films with wurtzite structure, the strain ε_{zz} can be evaluated by the biaxial strain model [38]. Along the *c*-axis, deduced from this model, the ε_{zz} component is calculated by the formula [39]:

$$\varepsilon_{zz} = \frac{c_{film} - c_{bulk}}{c_{bulk}} \tag{6.4}$$

where c_{bulk} is the lattice constant parameter along the *c*-axis of ZnO film without defect and is equal to 5.2066 Å [30], and c_{film} is the lattice constant parameter c of the film deduced from XRD measurements. According to Eq. (6.4), we calculated the strain in various doped ZnO thin films as shown in Table 6.1. The negative strain values ε_{zz} for TZO and AZO thin films reveal a compressive strain along the c-axis, while the positive value of ε_{zz} for CZO thin film is due to a tensile strain. Nevertheless, in each case, the low value of stain is linked to a slight structural defect and the high crystallization along the *c*-axis.

6.3.3 XRD Pole Figures Measurements

Figure 6.3 shows the X-ray pole figures of the hexagonal-doped ZnO thin films. Only (0002) planes have a markedly tendency to grow parallel to substrate surface. Therefore, Sn dopant is strongly *c*-axis oriented. These experimental results allow calculating, via the orientation function, the texture index of the films, which is found equal to 1.9, 5.5 and 4.5 for AZO, TZO, and CZO thin films, respectively.

6.3.4 Optical Properties

The transmittance $T(\lambda)$ spectra in the UV–VIS–NIR range (300–1100 nm) of the series of doped ZnO thin films are presented in Fig. 6.4. The transmittance spectra of all the films point out highly transparent films in the visible range with an average transmittance higher than 80% for TZO and AZO thin films but with an average transmittance equal to about 70% for CZO thin film. Nevertheless, this decrease of the transmittance in the visible range for CZO thin film is associated to a more abrupt UV absorption band edge shifting to shorter wavelengths than for other films.

The optical bandgap was determined from the absorption coefficient value. The energy gap (Eg) is considered assuming a direct transition between valence and conduction bands from the Tauc law expression [40, 41]:

$$(\alpha h\nu)^2 = A(h\nu - Eg) \tag{6.5}$$

where A is an energy-independent constant, $h\nu$ is the photon energy and Eg is the optical energy band and α is the absorption coefficient calculated from transmittance data [27]. We have plotted in Fig. 6.5 the characteristic $(\alpha h\nu)^2$ versus $h\nu$ for all the

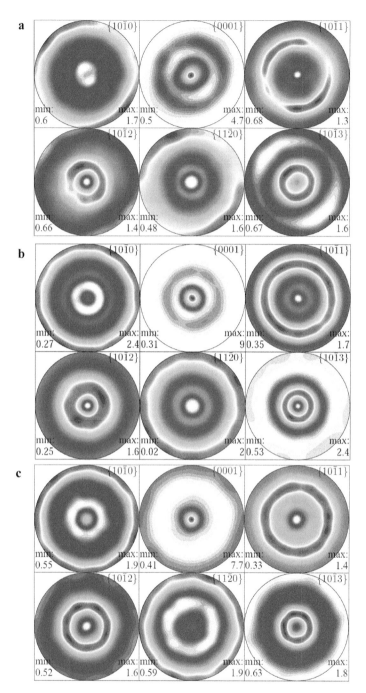

Fig. 6.3 X-ray pole figures of the hexagonal-doped ZnO thin films: **a** AZO, **b** TZO and **c** CZO

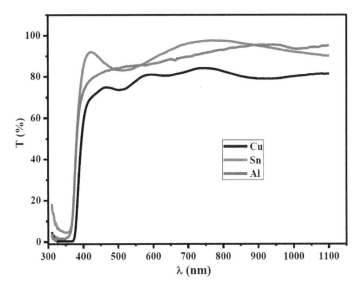

Fig. 6.4 Optical transmission spectra of Al, Sn, and Cu doped ZnO thin films

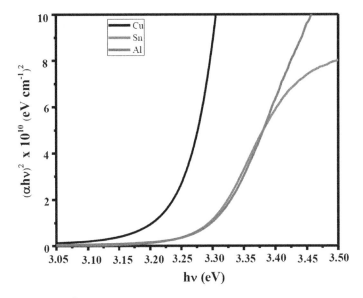

Fig. 6.5 Plots of $(\alpha h\nu)^2$ as function of the photon energy of Al, Sn, and Cu doped ZnO thin films

films. The optical band gaps values of the films, reported in Table 6.2, are determined by linear interpolation of the curves $(\alpha h\nu)^2 = 0$. The optical band gaps are found to be 3.30 eV, 3.28 eV and 3.27 eV for AZO, TZO and CZO thin films, respectively. The optical band gap of ZnO thin film changes with the doping.

Table 6.2 Calculated values of the optical bandgap Eg and Urbach energy Eu of the sprayed doped ZnO thin films

Thin films	Eg (eV)	Eu (eV)
AZO	3.30	91
TZO	3.28	101
CZO	3.27	98

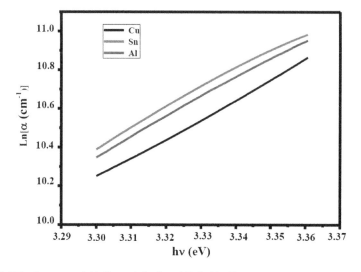

Fig. 6.6 Urbach energy of Al, Sn, and Cu doped ZnO thin films

The optical transitions between occupied states in the valence band tail to unoccupied states of the conduction band edge induce an exponential dependence of the absorption coefficient α on photon energy near the band edge [42]. To describe this phenomenon, the Urbach energy Eu is introduced. Eu refers to the width of the exponential absorption edge deduced from inverse of the slope of the curve $\ln(\alpha)$ as a function of $h\nu$, as shown in Fig. 6.6 and described by:

$$\alpha = \alpha_0 \exp\left(\frac{h\nu}{\text{Eu}}\right) \tag{6.6}$$

where α_0 is constant. For these doped ZnO thin films, the Urbach energy varies with doping and exhibits a minimum value of 91 meV for the AZO thin film (Table 6.2). These weak values indicting that the structural disorder is lower in the films which is confirmed the analysis done with XRD measurements.

6.4 Conclusion

Doped ZnO thin films with Al, Sn, and Cu have been deposited on glass substrates. They were elaborated by the simple and low-cost spray pyrolysis technique and were studied by techniques allowing characterizations of their structural, morphological and optical properties. The so-obtained thin films are polycrystalline with hexagonal wurtzite structure growing along the c-direction. TZO and AZO thin films reveal a compressive strain along the c-axis, while CZO thin film reveals tensile strain. Nevertheless, the low values of stain are linked to a slight structural defect and the high crystallization along the c-axis. The transmission properties of the films point out a transparency higher than 80% for TZO and AZO thin films but an average transmittance around 70% for CZO thin film in the visible range. The optical bandgap and the Urbach energy change with doping. The Urbach energy exhibits weak values of all the doped ZnO thin film corresponding to the signature of a low structural disorder in doped ZnO thin films which is confirmed the analysis done with XRD measurements. Finally, all characterizations confirm the high structural and optical quality of the so-obtained doped ZnO thin films by spray pyrolysis on glass substrate with hexagonal structure for optoelectronic devices.

References

1. Liu, C.Y., Zhang, B.P., Lu, Z.W., Binh, N.T., Wakatsuki, K., Segawa, Y., Mu, R.: Fabrication and characterization of ZnO film based UV photodetector. J. Mater. Sci.: Mater. Electron. **20**, 197–201 (2009). https://doi.org/10.1007/s10854-008-9698-x
2. Szarko, J.M., Song, J.K., Blackledge, C.W., Swart, I., Leone, S.R., Li, S., Zhao, Y.: optical injection probing of single zno tetrapod lasers. Chem. Phys. Lett. **404**, 171–176 (2005). https://doi.org/10.1016/j.cplett.2005.01.063
3. Ki Shin, B., Lee, T.I., Xiong, J., Hwang, C., Noh, G., Cho, J.H., Myoung, J.M.: Bottom-up grown ZnO nanorods for an antireflective moth-eye structure on CuInGaSe2 solar cells, Sol. Energy Mater. Sol. Cells, **95**, 2650–2654 (2011). https://doi.org/10.1016/j.solmat.2011.05.033
4. Peng, L., Xie, T.F., Yang, M., Wang, P., Xu, D., Pang, S., Wang, D.J.: Light induced enhancing gas sensitivity of copper-doped zinc oxide at room temperature. Sens. Actuators B: Chem. **131**, 660–664 (2008). https://doi.org/10.1016/j.snb.2007.12.060
5. Wahab, R., Kaushik, N.K., Verma, A.K., Mishra, A., Hwang, I.H., Yang, Y.B., Shin, H.S., Kim, Y.S.: Fabrication and growth mechanism of ZnO nanostructures and their cytotoxic effect on human brain tumor U87, cervical cancer HeLa, and normal HEK cells. J. Biol. Inorg. Chem. **16**, 431–442 (2011). https://doi.org/10.1007/s00775-010-0740-0
6. Wang, Z.L.: Zinc oxide nanostructures: growth, properties and applications J. Phys.: Condens. Matter. **16**, R829–R858 (2004). https://doi.org/10.1088/0953-8984/16/25/R01
7. Yang, J., Lee, J., Kim, K., Lim, S.: Influence of Sn-doping in hydrothermal methods on the optical property of the ZnO nanorods. Physi. E **42**, 51–56 (2009). https://doi.org/10.1016/j.physe.2009.08.018
8. Ozgur, U., Alivov, Ya.I., Liu, C., Teke, A., Reshchikov, M.A., Doan, S., Avrutin, V., Cho, S.-J., Morkoc, H.: A comprehensive review of ZnO materials and devices. J. Appl. Phys. **98**, 041301 (2005). https://doi.org/10.1063/1.1992666

9. Zamfirescu, M., Kavokin, A., Gil, B., Malpuech, G., Kaliteevski, M.: ZnO as a material mostly adapted for the realization of room-temperature polariton lasers. Phys. Rev. B **65**, 161205 (2002). https://doi.org/10.1103/PhysRevB.65.161205
10. Saravanakumar, K., Ravichandran, K., Chandramohan, R., Gobalakrishnan, S.: Investigation on simultaneous doping of Sn and F with ZnO nanopowders synthesized using a simple soft chemical route. Superlattice. Microstruct. **52**, 528–540 (2012). https://doi.org/10.1016/j.spmi. 2012.06.003
11. Kim, ChE, Moon, P., Kim, S., Myoung, J.M., Jang, H.W., Bang, J., Yun, I.: Effect of carrier concentration on optical bandgap shift in ZnO: Ga thin films. Thin Solid Films **518**, 6304–6307 (2010). https://doi.org/10.1016/j.tsf.2010.03.042
12. Mariappan, R., Ponnuswamy, V., Suresh, P.: Effect of doping concentration on the structural and optical properties of pure and tin doped zinc oxide thin films by nebulizer spray pyrolysis (NSP) technique. Superlattice. Microstruct. **52**, 500–513 (2012). https://doi.org/10.1016/j.spmi.2012. 05.016
13. Bouderbala, M., Hamzaoui, S., Amrani, B., Reshak, A.H., Adnane, M., Sahraoui, T., Zerdali, M.: Thickness dependence of structural, electrical and optical behaviour of undoped ZnO thin films. Phys. B **403**, 3326–3330 (2008). https://doi.org/10.1016/j.physb.2008.04.045
14. Bedia, A., Bedia, F.Z., Aillerie, M., Maloufi, N., Ould Saad Hamady, S., Perroud, O., Benyoucef, B.: Optical, electrical and structural properties of nano-pyramidal ZnO films grown on glass substrate by spray pyrolysis technique. Opt. Mater. **36**(7), 1123–1130 (2014). https://doi.org/ 10.1016/j.optmat.2014.02.012
15. Samanta, P.K., Basak, S.: Electrochemical growth of hexagonal ZnO pyramids and their optical property. Mater. Lett. **83**, 97–99 (2012). https://doi.org/10.1016/j.matlet.2012.05.133
16. Jagadish, C., Pearton, S.: Zinc Oxide Bulk, 1st ed., Thin Films Nanostructures, Elsevier (2006)
17. Tian, Y., Lu, H.B., Li, J.C., Wu, Y., Fu, Q.: Synthesis, characterization and photoluminescence properties of ZnO hexagonal pyramids by the thermal evaporation method Physi. E **43**, 410–414 (2010). https://doi.org/10.1016/j.physe.2010.08.024
18. Bedia, A., Bedia, F.Z., Aillerie, M., Maloufi, N.: Structural, electrical and optical properties of Al–Sn codoped ZnO transparent conducting layer deposited by spray pyrolysis technique. Superlattice. Microstruct. **111**, 714–721 (2017). https://doi.org/10.1016/j.spmi.2017.07.031
19. Janisch, R., Gopal, P., Spaldin, N.A.: Transition metal-doped TiO$_2$ and ZnO—present status of the field. J. Phys.: Condens. Matter **17**, R657–R689 (2005). https://doi.org/10.1088/0953-8984/17/27/R01
20. Paraguay, F., Miki-Yoshida, D.M., Morales, J.: Influence of Al, In, Cu, Fe and Sn dopants on the response of thin film ZnO gas sensor to ethanol vapour. J. Solis, W. Estrada L., Thin Solid Films **73**, 137 (2000). https://doi.org/10.1016/S0040-6090(00)01120-2
21. Han, N., Wu, X., Zhang, D., Shen, G., Liu, H., Shen, Y.: CdO activated Sn-doped ZnO for highly sensitive, selective and stable formaldehyde sensor. Sens. Act. B: Chem. **152**, 324–329 (2011). https://doi.org/10.1016/j.snb.2010.12.029
22. Farag, A.A.M., Cavas, M., Yakuphanoglu, F., Amanullah, F.M.: Photoluminescence and optical properties of nanostructure Ni doped ZnO thin films prepared by sol–gel spin coating technique. J. Alloys Comp. **509**, 7900–7908 (2011). https://doi.org/10.1016/j.jallcom.2011.05.009
23. Mhamdi, A., Boukhachem, A., Madani, M., Lachheb, H., Boubaker, K., Amlouk, A., Amlouk, M.: Study of vanadium doping effects on structural, opto-thermal and optical properties of sprayed ZnO semiconductor layers. Optik **124**, 3764–3770 (2013). https://doi.org/10.1016/j. ijleo.2012.11.074
24. Boukhachem, A., Ouni, B., Karyaoui, M., Madani, A., Chtourou, R., Amlouk, M.: Structural, opto-thermal and electrical properties of ZnO: Mo sprayed thin films. Mater. Sci. Semicond. Process. **15**, 282–292 (2012). https://doi.org/10.1016/j.mssp.2012.02.014
25. Ilican, S., Caglar, Y., Caglar, M., Yakuphanoglu, F.: Structural, optical and electrical properties of F-doped ZnO nanorod semiconductor thin films deposited by sol–gel process Appl. Surf. Sci. **255**, 2353–2359 (2008). https://doi.org/10.1016/j.apsusc.2008.07.111
26. Caglar, M., Yakuphanoglu, F.: Structural and optical properties of copper doped ZnO films derived by sol–gel. Appl. Surf. Sci. **258**, 3039–3044 (2012). https://doi.org/10.1016/j.apsusc. 2011.11.033

27. Bedia, F.Z., Bedia, A., Maloufi, N., Aillerie, M., Genty, F., Benyoucef, B.: Effect of tin doping on optical properties of nanostructured ZnO thin films grown by spray pyrolysis technique. J. Alloys Compd. **616**, 312–318 (2014). https://doi.org/10.1016/j.jallcom.2014.07.086

28. Swanepoel, R.: Determination of the thickness and optical constants of amorphous silicon. J. Phys. E: Sci. Instrum. **16**, 1214–1222 (1983). https://doi.org/10.1088/0022-3735/16/12/023

29. Hielscher, R., Schaeben, H.: A novel pole figure inversion method: specification of the MTEX algorithm. J. Appl. Cryst. **41**, 1024–1037 (2008). https://doi.org/10.1107/S0021889808030112

30. JCPDS Card No. 36–1451 (ZnO hexagonal)

31. Rao, T.P., Santhoshkumar, M.C., Safarulla, A., Ganesan, V., Barman, S.R., Sanjeeviraja, C.: Physical properties of ZnO thin films deposited at various substrate temperatures using spray pyrolysis. Phys. B **405**, 2226 (2010). https://doi.org/10.1016/j.physb.2010.02.016

32. Singhal, S., Kaur, J., Namgyal, T., Sharma, R.: Cu-doped ZnO nanoparticles: synthesis, structural and electrical properties. Phys. B **407**, 1223–1226 (2012). https://doi.org/10.1016/j.physb.2012.01.103

33. Ilican, S., Caglar, M., Caglar, Y.: Sn doping effects on the electro-optical properties of sol gel derived transparent ZnO films. Appl. Surf. Sci. **256**, 7204–7210 (2010). https://doi.org/10.1016/j.apsusc.2010.05.052

34. Yoo, R., Cho, S., Song, M.J., Lee, W.: Highly sensitive gas sensor based on Al-doped ZnO nanoparticles for detection of dimethyl methylphosphonate as a chemical warfare agent stimulant. Sens. Actuat. B **221**, 217–223 (2015). https://doi.org/10.1016/j.snb.2015.06.076

35. Ma, L., Ma, S., Chen, H., Ai, X., Huang, X.: Microstructures and optical properties of Cu-doped ZnO films prepared by radiofrequency reactive magnetron sputtering. Appl. Surf. Sci. **257**, 10036–10041 (2011). https://doi.org/10.1016/j.apsusc.2011.06.134

36. Swapna, R., Santhosh Kumar, M.C.: Growth and characterization of molybdenum doped ZnO thin films by spray pyrolysis. J. Phys. Chem. Solids **74**, 418–425 (2013). https://doi.org/10.1016/j.jpcs.2012.11.003

37. Djelloul, A., Aida, M.S., Bougdira, J.: Photoluminescence, FTIR and X-ray diffraction studies on undoped and Al-doped ZnO thin films grown on polycrystalline α-alumina substrates by ultrasonic spray pyrolysis. J. Lumin. **130**, 2113–211721 (2010). https://doi.org/10.1016/j.jlumin.2010.06.002

38. Segmuller, A., Murakami, M.: In: Tu, K.N., Rosenberg, R. (eds.) Analytical Techniques Thin Films, p. 143. Academic, Boston (1988)

39. Bahedi, K., Addoua, M., El Jouada, M., Sofiania, Z., EL Oauzzanib, H., Sahraouib, B.: Influence of strain/stress on the nonlinear-optical properties of sprayed deposited ZnO: Al thin films. Appl. Surf. Sci. **257**, 8003–8005 (2011). https://doi.org/10.1016/j.apsusc.2011.04.072

40. Pankove, J.I.: Optical processes in semiconductors. Prentice-Hall Inc., Englewood Cliffs, NJ (1971)

41. Tauc, J., Grigorovici, R., Vancu, A.: Optical properties and electronic structure of amorphous germanium. Phys. Stat. Sol. **15**, 627–637 (1966). https://doi.org/10.1002/pssb.19660150224

42. Urbach, F.: The long-wavelength edge of photographic sensitivity and of the electronic absorption of solids. Phys. Rev. **92**, 1324 (1953). https://doi.org/10.1103/PhysRev.92.1324

Part II
PHOTOVOLTAIC SYSTEMS: Methods and Applications

Chapter 7
New Reconfiguration Method Based on Logic Gates for Small Dynamic Photovoltaic Array

L. Bouselham, A. Rabhi, and B. Hajji

Abstract This chapter proposes a new method for reconfiguring the dynamic photovoltaic (PV) array under repeating shade conditions. The repeating shades are often caused in photovoltaic installations, especially in residential installations where PV modules can be subjected to shades occurred by nearby buildings or trees. The proposed method is based on logic gates and aims to minimize the processing time in the way that controller does not have to perform an exhaustive calculations at each shade condition to achieve the optimal configuration of the PV generator . Simulation of 2×2 size dynamic photovoltaic array has been carried out. Experimental tests of 1 $\times$ 1 size Dynamic Photovoltaic array under different irradiance conditions have been also conducted. The simulation and experimental tests have validated the proposed method in identification of the optimal configuration with less processing time and with an improvement in reducing power losses.

Keywords Dynamic PV array · Switches network · Irradiance equalization · Logic gates

7.1 Introduction

The rising demand for electricity, the depletion of conventional energy resources and the growing concern for environmental issues have favoured the use of renewable energy sources for electricity generation. Particularly, solar energy is acquiring much acclaim due to the growth of nano-semiconductor technology with the fast declining cost of photovoltaic (PV) cells area [1]. In addition, a PV system does not cause

L. Bouselham · B. Hajji (✉)
Renewable Energy, Embedded System and Data Processing Laboratory, National School of
Applied Sciences, Mohamed First University, Oujda, Morocco
e-mail: hajji.bekkay@gmail.com

A. Rabhi
Modelization, Information and Systems Laboratory, Picardie Jules Verne University, Amiens,
France

A. Mellit and M. Benghanem (eds.), *A Practical Guide for Advanced Methods
in Solar Photovoltaic Systems*, Advanced Structured Materials 128,
https://doi.org/10.1007/978-3-030-43473-1_7

noise, which promotes its implementation in the urban area, as well as it requires less maintenance [2]. However, despite its many advantages, the power generated by a PV module is limited. The output power is nonlinear and highly dependent on environmental conditions.

Due to the nonlinear output characteristics of the PV module, tracking the maximum power point (MPP) in various environmental conditions can be sometimes a challenging task. The issue becomes more delicate when it is a PV generator (PVG) with several PV modules connected to each other and the whole PVG does not receive a uniform irradiance [3]. This situation often occurs with partial shading (PS).

PS is a frequent phenomenon which reduces considerably the output power of a PV system. It can be caused by moving clouds, snow, tree shadow or bird dung covering the surface of PV modules. When PS occurs, the shaded modules start to produce less current, while the unshaded modules continue to operate at a higher current. Therefore, the shaded modules will impose a current limitation in the string. In addition, if they are forced to drive more current, they will be polarized in reverse, resulting in hot spots [4].

The hot spot is an undesirable effect because it can permanently destroy the shaded PV modules. So, the bypass diodes are connected in parallel to the PV modules in the way to divert the current through the path provided by the bypass diode [5]. However, the power–voltage (P-V) characteristic of PVG may heavily be influenced. It exhibits multiple local maxima (LMs), and only one of them corresponds to the global maximum (GM) [6]. In this case, if the MPPT technique pursues one of the LMs instead of the GM, then the estimated power loss can go up to 70% [7].

Since the performance of a GPV under partial shading varies depending on the location of the shaded modules in the GPV as well as the interconnection scheme of the PV modules, a dynamic reconfiguration of PV module connections according to the environment changes can prevent the decrease of the output power [8]. The PV modules are rearranged so that the unshaded PV modules compensate the shaded modules for a smoothest P-V characteristic with only one MPP, thus avoiding the use of complex MPPT algorithm. The PVG reconfiguration will be achieved by integrating a switching network (SN) into the PVG. The SM is controlled by an optimization algorithm which identifies the best configuration of the PVG at each irradiance-level variation (Fig. 7.1).

In literature, many dynamic PV array (DPVA) approaches have been reported:

- Adaptive PV array with fixed part [9, 10]: a small adaptive part of the PV modules is connected to a fixed part of the PV modules through the SM. At each PS occurrence, the high current row of the adaptive part is connected to the shaded row with the low current of the fixed part. This approach requires simple SM with less number of switches since only the modules of the dynamic part are relocated. However, to effectively deal with all possible shading scenarios, a huge adaptive bank with a large number of sensors is required.
- Irradiance categorization approach [11, 12]: the PV modules are classified according to their irradiance value (clear category and dark category). The dark category is excluded from the PVG, since their power contribution is negligible.

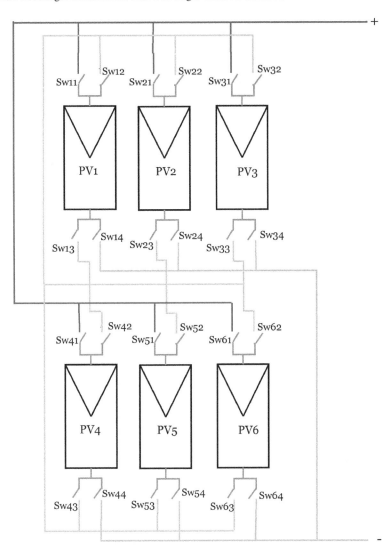

Fig. 7.1 Example of dynamic PV array

- Logical approach [13–17]: the optimal PVG configuration is identified by using intelligent algorithm like genetic algorithm and particle swarm optimization. Despite the reduced computational time insured by logical approach, however, it requires an intense parameter to be tuned for an optimal performance.
- Irradiance equalization (IEq) approach: applicable to PV modules that are connected in total-cross-tied (TCT) and the most opted reconfiguration approach for

DPVA. Depending on the irradiance received by each module, the PV modules are moved such that each row in the series string has the same amount of irradiance (discussed in Sect. 7.2).

This chapter develops research for the last approach. A new adaptive reconfiguration of solar panels based on the IEq is proposed. The developed method controls the matrix of the switches by logic gates (LG), in order to ensure the optimal configuration of the PVG. Some of the main contributions of this work are summarized below:

- Simple method of reconfiguration applied to any small-scale PV array is proposed, able to follow the fast variations of shading pattern by minimizing the treatment time to achieve the optimal configuration.
- Small-scale outdoor tests under various shading scenarios are conducted. Results have validated the performance of the proposed DPVA.

The remaining parts of the chapter are organized as follows. Section 7.2 provides a brief review on IEq methods and the problem statement. Description of the proposed control method with LG is presented in Sect. 7.3. Section 7.4 summarizes the simulation results under MATLAB Simulink. Experimental set-up is described in Sect. 7.5. Finally, Sect. 7.6 outlines the main conclusions.

7.2 Irradiance Equalization Approach Principal and Problem Statement

The IEq approach introduced in [18] aims to form rows with similar average irradiance values or sufficiently close to each other, depending on the existing irradiance profile (Fig. 7.2). The optimization algorithm selects the configuration which minimizes the IE. Thus, by examining all possible PVG configurations, an important processing time is needed.

The reconfiguration of PV modules is insured by a random search algorithm and a deterministic search algorithm in [19] with the capability of reconfiguration with non-equal module number per row. For an optimal choice of the appropriate configuration, authors in [20] propose Smart Choice (SC) algorithm based on the Munkres' assignment associated with dynamic programming (DP). Both DP and SC algorithms seek to resolve the IEq problem. The suggested configurations by the two algorithms are compared and that minimizing the IE and the number of switching operations N_{SW} is maintained. However, the simultaneous execution of the two algorithms requires a very important processing time. Authors in [21] propose an iterative and hierarchical sorting algorithm based on IEq method. An improved sorting algorithm is also proposed [22] in order to obtain the appropriate configuration with less computation time and taking into account the number of N_{SW}. However, in some cases, just near-optimal configuration with quasi-equalized rows is achieved.

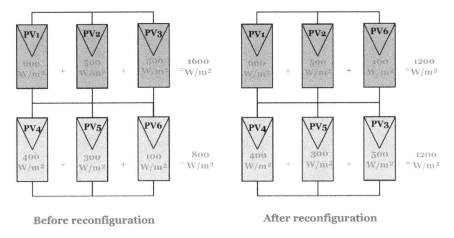

Before reconfiguration   **After reconfiguration**

Fig. 7.2 Irradiance equalization principal

The sum of irradiance on each row after reconfiguration is not equal but close to the average level (Fig. 7.3):

$$\text{Avg} = \frac{\sum_{i=1}^{m} G_i}{m} \tag{1}$$

where m is the number of rows.

A configuration with quasi-equalized irradiance rows causes multiple peaks in P-V characteristic as a fixed TCT interconnection as shown in Fig. 7.4.

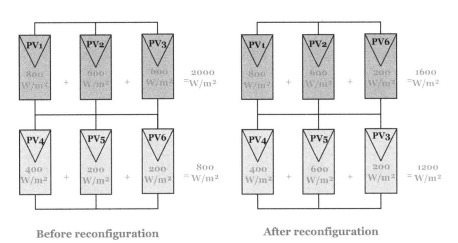

Before reconfiguration   **After reconfiguration**

Fig. 7.3 Example of quazi-equalized irradiance rows

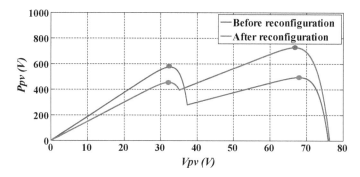

Fig. 7.4 P-V characteristic before and after reconfiguration in the case of quazi-equalized irradiance rows

7.3 Proposed Reconfiguration Method

The proposed method aims to achieve the optimal configuration with less treatment time by using a LG instead of the optimization algorithm. In addition, the LG will be activated in the way to ensure equalized rows in irradiance by keeping the same number of PV modules by rows or by forming PVG with non-equal number of PV modules by rows (Fig. 7.5).

For understanding purposes, we consider a GPV of size 2×2, interconnected on TCT topology, throw SN of 24 single pole, single throw (SPST) switches. Each switch is controlled by a set of LGs that receive as inputs the irradiation states (G) of the PV panels (Fig. 7.6). Since LG admits only two states (0 or 1), irradiation is categorized into two levels:

- Light level: $G \geq 500 \, \text{W/m}^2$ corresponding to state "1"
- Dark level: $G \geq 500 \, \text{W/m}^2$ corresponding to state "0"

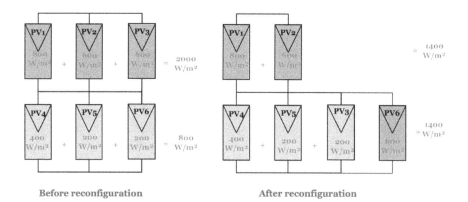

Before reconfiguration   After reconfiguration

Fig. 7.5 Principal of proposed reconfiguration method

Fig. 7.6 Example of logic gates reconfiguration command

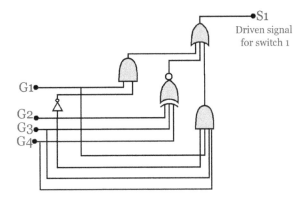

Truth table is used to identify the driven signals (S) of switches for all possible combinations of $\{G_1, G_2, G_3; G_4\}$ (Table 7.1). For each combination, the switches are activated in the way to ensure equalized rows in irradiance as shown in the example depicted in Fig. 7.7.

The logical expression of each output to activate the switches is deduced using the Karnaugh chart as follows:

$$S_{11} = S_{13} = \overline{S_{12}} = \overline{S_{14}} = G_1.\overline{G_2} + \overline{G_2 \oplus G_3 \oplus G_4} + G_1 \cdot G_2 \cdot G_3 \cdot G_4$$
$$S_{21} = S_{23} = \overline{S_{22}} = \overline{S_{24}} = \overline{G_1} \cdot \overline{G_3} \cdot \overline{G_4} + G_2 + G_1 \cdot (G_3 \oplus G_4)$$
$$S_{31} = S_{33} = \overline{S_{32}} = \overline{S_{34}} = \overline{G_1} \cdot \overline{G_2} \cdot G_3 + G_1 \cdot G_3(G_2 \oplus G_4) + \overline{G_1} \cdot G_3 \cdot G_4$$
$$S_{41} = S_{43} = \overline{S_{42}} = \overline{S_{44}} = G_1 \cdot G_2 \cdot G_4 + \overline{G_1} \cdot \overline{G_3} \cdot \overline{G_2} \cdot G_4 + G_4 \cdot G_3 \cdot (G_1 \oplus G_2)$$

7.4 Simulations for 2 × 2 Size Dynamic PV Array

To check the effectiveness of the proposed logic gates reconfiguration method, MAT-LAB simulations of 2 × 2 size DPVA have been conducted. The Perturb and Observe (P&O) algorithm is used as MPPT. The simulated diagram is shown in Fig. 7.8. For comparison purpose, the basic IEq algorithm (B-IEq) and the sorting algorithm that are already developed (S-IEq) are used.

Simulation response of the proposed DPVA under the repeating PS conditions is shown in Fig. 7.9. In total simulation time of 80 ms, from 0 to 20 ms is the DPVA response to the shade-1 test, 20–40 ms is the response to shade-2 test, 40–60 ms is the response to shade-3 test and 60–80 ms is the response to the shade-4 test.

During shade-1 test, the PVG produces 681 W with the proposed method however the PVG output power has been reduced to 453 W. This result is justified by that the proposed method success to configure the PVG with equalized rows or rows with

Table 7.1 Truth table for 2 × 2 GPV

G_1	G_2	G_3	G_4	S_{11}	S_{12}	S_{13}	S_{14}	S_{21}	S_{22}	S_{23}	S_{24}	S_{31}	S_{32}	S_{33}	S_{34}	S_{41}	S_{42}	S_{43}	S_{44}
0	0	0	0	1	0	1	0	1	0	1	0	0	1	0	1	0	1	0	1
0	0	0	1	0	1	0	1	1	0	1	0	1	0	1	1	0	1	0	0
0	0	1	0	0	1	0	1	1	0	1	1	0	1	0	0	1	0	1	1
0	0	1	1	1	0	1	0	1	0	1	1	0	1	0	0	1	0	1	1
0	1	0	0	0	1	0	1	0	1	0	0	1	0	1	0	1	0	1	1
0	1	0	1	1	0	1	0	0	1	0		1	0	1	0	1	0	1	1
0	1	1	0	1	0	1	0	0	1	0	0	1	0	1	0	0	0	0	1
0	1	1	1	0	1	0	1	0	1	0	1	0	1	0	1	1	1	1	0
1	0	0	0	1	0	1	0	1	0	1	0	1	0	1	0	1	0	1	1
1	0	0	1	1	0	1	0	0	1	0	0	1	0	1	0	1	0	1	1
1	0	1	0	1	0	1	0	0	1	0	0	1	0	1	0	0	0	0	1
1	0	1	1	0	0	0	0	1	0	1	1	0	1	0	1	0	1	0	0
1	1	0	0	1	1	1	1	0	1	0	0	1	0	1	1	0	1	0	0
1	1	0	1	1	0	1	0	0	1	0	0	1	0	1	1	0	1	0	0
1	1	1	0	1	0	1	0	1	0	1	0	1	0	1	0	0	1	0	1
1	1	1	1	1	0	1	0	1	0	1	0	0	1	0	1	0	1	0	1

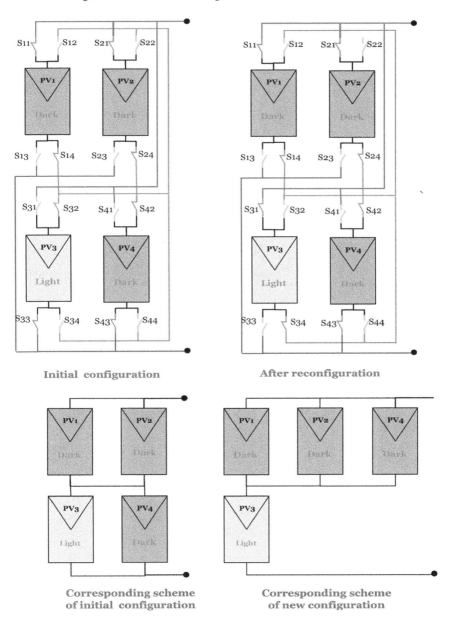

Initial configuration

After reconfiguration

Corresponding scheme
of initial configuration

Corresponding scheme
of new configuration

Fig. 7.7 Example of switches stats before and after reconfiguration and their corresponding scheme

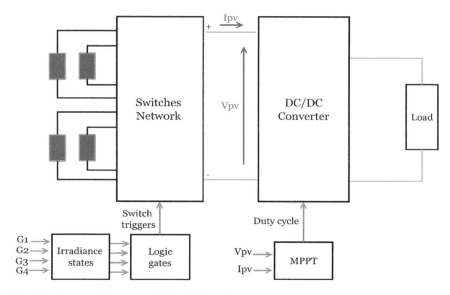

Fig. 7.8 Simulated diagram in MATLAB/Simulink

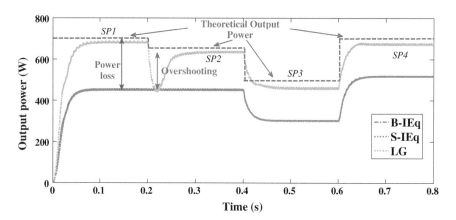

Fig. 7.9 Response of proposed DPVA scheme under repeating PS conditions

current values close to each other, whereas the B-IEq and S-IEq methods fall into the quazi-equalization situation, where 48% of power is lost.

At 20 ms, the B-IEq and S-IEq methods keep the same previous power value since just one PV module is shaded for this shade test. However, the PVG output power had been increased to 635 W with the proposed method. Total time taken is 10 ms to identify the reconfiguration solution and for the MPPT to track the MPP. An overshoot is observed from shade-1 test to shade-2 test due to the duty cycle search of its new value.

For the shade-3 and shade-4 tests, the same remarks as the shade-1 can be made. Compared to the B-IEq and S-IEq methods, an improvement of power loss is achieved by the proposed method estimated at 18%.

7.5 Experimental Test of 1 × 1 Size Dynamic PV Array

To demonstrate the feasibility of a DPVA as well as the proposed reconfiguration method, simple experimental tests of 1 × 1 size DPVA are conducted. The GPV is made up of two PV panels which they are connected to each other in series or in parallel depending on the irradiation value received by each module:

- If $G_1 = G_2$, the PV panels are connected in series
- If $G_1 \neq G_2$, the PV panels are connected in parallel

The switching from one connection to the other is ensured by three relays as shown in Fig. 7.10. The Truth table which identifies the driven signals of relays for

Fig. 7.10 Block diagram of experienced GPV

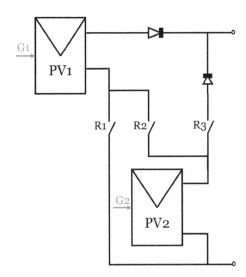

Table 7.2 Truth table for 1 × 1 GPV

G_1	G_2	R_1	R_2	R_3
0	0	1	0	1
0	1	0	1	0
1	0	0	1	0
1	1	1	0	1

all possible combinations of $\{G_1, G_2\}$ is depicted in Table 7.2. The logical expression of each driven signal (R) for relays is deduced as follows:

$$R_1 = R_3 = \overline{R_2} = \overline{G_1 \oplus G_2}$$

Photographic view of the test set-up is shown in Fig. 7.11. A brief description of test set-up components is given in Table 7.3.

As shown in Fig. 7.11, the Dspace card is used as a controller. It collects data from each sensor, identifies the best solution and sends high/low logic signals to control

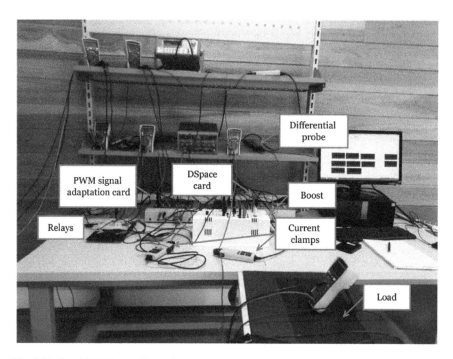

Fig. 7.11 1×1 DPVA experimental test set-up

Table 7.3 Description of experimental test components

Component	Specifications
PVG	2 PV modules, each rated for 300 W
Boost converter	$C_e = 330\,\mu F$, $L = 1\,mH$, $C_s = 1100\,\mu F$
Sensors	Differential probes GE 8115 Current clamps i30s
Controller	Dspace CLP1104
SN	3 Relays HF115F-Q
Load	Resistive load 25 Ω

the relays. In addition, it defines the appropriate value of the duty cycle according to P&O MPPT. The used PV modules are installed in the roof of a chalet.

The robustness of the PVG to switch from one configuration to another is tested by creating artificial shading on one of the PV panels at different time intervals (Fig. 7.12).

Figure 7.13 describes the behaviour of the three relays. With the existence of shading, the architecture is configured in parallel. In this case, the relays R1 and

Fig. 7.12 Artificial shading created on the PV panels

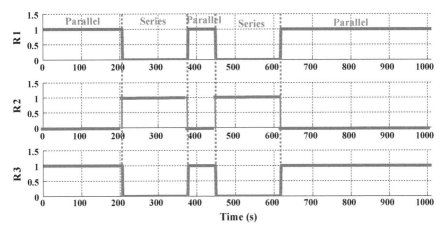

Fig. 7.13 Relays behaviour

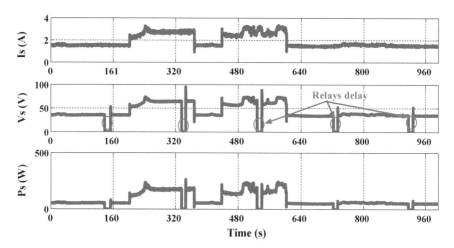

Fig. 7.14 Current, voltage and output power profiles

R3 are closed. Otherwise, the architecture is configured in series and only the relay R2 which is closed. With the transition from a parallel configuration to a serial configuration, the current increases.

As shown in Fig. 7.14, with the transition from a parallel configuration to a serial configuration, the current increases. As a transition from a serial configuration to a parallel configuration, the current decreases. The same situation applies to the voltage, except that the voltage is affected by the response time of the relays. A delay is observed in the voltage response when switching from one configuration to another.

To sum up, at each shade test, the controller is able to detect PS condition, identify the reconfiguration solution and stabilize quickly the MPPT. In fact, fast processing time ensures the good monitoring of fast repeating PS conditions. However, the treatment time depends on the speed of the processor chosen for the task. Moreover, the identification of the reconfiguration solution depends in addition to the processing time of the controller, to the propagation time of the LGs which is generally very fast.

7.6   Conclusion

This chapter proposes a new DPVA reconfiguration method based on logic gates. The aim of this work is to avoid the quasi-equalized rows in irradiance after reconfiguration. The proposed method reconfigures the GPV with the same number of PV panels per row or by an unequal number of PV panels per row. Moreover, the proposed method uses logic gates to activate the switches in order to minimize the processing time and thus ensure the new optimal configuration even if rapid variations of shade have occurred. MATLAB simulations of 2-by-2 DPVA are carried out. A significant improvement in reduction of power losses is achieved, estimated at 18% compared to the basic IEq algorithm and the sorting IEq already proposed. Further, experimental tests of 1×1 size DPVA are conducted. Results validate the effectiveness of the proposed method to identify the optimal configuration under variable irradiance conditions. However, the proposed method should be applied to small PV systems. As the inputs of logic gates depend on irradiance state of each PV panel, the identification of the logical expression of each driven signal for switches in a large PV scale becomes complicated task.

References

1. Mutoh, N., Ohno, M., Inoue, T.: A method for MPPT control while searching for parameters corresponding to weather conditions for PV generation systems. IEEE Trans. Ind. Electron. **53**(4), 1055–1065 (2008)
2. Khezzar, R., Zereg, M., Khezzar, A.: Modeling improvement of the four parameter model for photovoltaic modules. Sol. Energy **110**, 452–462 (2014)
3. Ishaque, K., Syafaruddin, Salam Z.: A comprehensive MATLAB Simulink PV system simulator with partial shading capability based on two-diode model. Sol. Energy **85**(9), 2217–2227 (2011)
4. Bidram, A., Davoudi, A., Balog, R.S.: Control and circuit techniques to mitigate partial shading effects in photovoltaic arrays. IEEE J. Photovolt. **2**(4), 532–546 (2012)
5. Malathy, S., Ramaprabha, R., Mathur, B.L.: Asymmetrical multilevel inverters for partially shaded PV systems. In: Proceedings of IEEE Conference on Circuits, Power and Computing Technologies, pp. 579–58 (2013)
6. Jaleel, J.A., Omega, A.R.: Maximum power point tacking simulation of PV array at partially shaded condition using lab view. In: Proceedings IEEE Conference on Control Communication and Computing, pp. 319–324 (2013)
7. Koutroulis, E., Blaabjerg, F.: A new technique for tracking the global maximum power point of PV arrays operating under partial-shading conditions. IEEE J. Photovolt. **2**(2), 184–190 (2012)
8. Malathy, S., Ramaprabha, R.: Reconfiguration strategies to extract maximum power from photovoltaic array under partially shaded conditions. Renew. Sustain. Energy Rev. 1–13 (2017)
9. Nguyen, D., Member, S., Lehman, B.: An adaptive solar photovoltaic array using model-based reconfiguration algorithm. IEEE Trans. Ind. Electron. **55**(7), 2644–2654 (2008)
10. Nguyen, D,, Lehman, B.: A reconfigurable solar photovoltaic array under shadow conditions. In: IEEE Twenty-Third Annual Applied Power Electronics Conference and Exposition, pp. 980–986 (2008)
11. Patnaik, B., Sharma, P., Trimurthulu, E., Duttagupta, S.P., Agarwal, V.: Reconfiguration strategy for optimization of solar photovoltaic array under non-uniform illumination conditions. In: IEEE 37th Photovoltaic Specialists Conference, pp. 1859–1864 (2011)
12. Patnaik, B., Mohod, J.D., Duttagupta, S.P.: Distributed multi-sensor network for real time monitoring of illumination states for a reconfigurable solar photovoltaic array. In: IR International Symposium on Physics and Technology of Sensors, pp. 106–109 (2012)
13. Shams El-Dein, M.Z., Kazerani, M., Salama, M.M.A.: Optimal photovoltaic array reconfiguration to reduce partial shading losses. IEEE Trans. Sustain. Energy **4**, 145–153 (2013). https://doi.org/10.1109/TSTE.2012.2208128
14. Orozco-Gutierrez, M.L., Spagnuol, G., Ramirez-Scarpetta, J.M., Petrone, G., Ramos-Paja, C.A.: Optimized configuration of mismatched photovoltaic arrays. IEEE J. Photovolt. **6**, 1210–1220 (2016). https://doi.org/10.1109/JPHOTOV.2016.2581481

15. Carotenuto, P.L., Della Cioppa, A., Marcelli, A., Spagnuolo, G.: An evolutionary approach to the dynamical reconfiguration of photovoltaic fields. Neurocomputing **170**, 393–405 (2015)
16. Rajan, N.A., Shrikant, K.D., Dhanalakshmi, B., Rajasekar, N.: Solar PV array reconfiguration using the concept of standard deviation and genetic algorithm. Energy Procedia **117**, 1062–1069 (2017)
17. Babu, T.S., Ram, J.P., Dragicevic, T., Miyatake, M., Blaabjerg, F., Rajasekar, N.: Particle swarm optimization based solar PV array reconfiguration of the maximum power extraction under partial shading conditions. IEEE Trans. Sustain. Energy **9**, 74–85 (2018). https://doi.org/10.1109/TSTE.2017.2714905
18. Velasco, G., Guinjoan-gispert, F., Piqué-López, R., Román-lumbreras, M., Conesa-roca, A.: Electrical PV array reconfiguration strategy for energy extraction improvement in grid-connected PV systems. IEEE Trans. Ind. Electron. **56**(11), 4319–4331 (2009)
19. Romano, P., Candela, R., Cardinale, M., Vigni V, L., Sanseverino, E.R.: Optimization of photovoltaic energy production through an efficient switching matrix. J. Sustain. Dev. Energy, Water Environ. Syst. **1**(3), 227–236 (2013)
20. Ngo Ngoc, T., Phung, Q.N., Tung, L.N., Riva Sanseverino, E., Romano, P., Viola, F.: Increasing efficiency of photovoltaic systems under non-homogeneous solar irradiation using improved dynamic programming methods. Sol. Energy **150**, 325–334 (2017)
21. Wilson, P., Storey, J., Bagnall, D.: Improved optimization strategy for irradiance equalization in dynamic photovoltaic arrays. IEEE Trans. Power Electron. **28**, 2946–2956 (2013)
22. Bouselham, L., Hajji, B., Mellit, A., Rabhi, A.: A reconfigurable PV architecture based on new irradiance equalization algorithm. In: Lecture Notes in Electrical Engineering, pp. 470–477 (2018)

Chapter 8
Energy Storage and Photovoltaic Systems

S. Blaifi

Abstract The storage in renewable energy systems especially in photovoltaic systems is still a major issue related to their unpredictable and complex working. Due to the continuous changes of the source outputs, several problems can be encountered for the sake of modeling, monitoring, control and lifetime extending of the storage devices. Therefore, several storage devices were introduced in the practice such as pumped hydro, compressed air, flywheel, super capacitors and electrochemical storage. However, the electrochemical storage especially the storage by battery bank is still the most used in PV systems. According to the performances and the features needed in such systems, two batteries types can be distinguished, namely lithium-ion and lead-acid-based batteries. Likely, there is a consensus that the lithium battery presents a better performances comparing to other types such as the high energy density, the low self-discharge current and the low maintenance. However, the major disadvantage of these batteries type is their high-cost which somewhat has slow down their progress for the large-scale applications. From there, the storage using lead-acid battery type is still widely used for the reason of its low cost and the ease of its maintenance. However, its complex electrochemical and electrical behaviors besides the random working of the PV systems make it as one of main issues for the sake of modeling, control and lifetime extending. In this chapter, we provide description of dynamic batteries behavior, encountered problems in the PV systems with solutions proposal in terms of modeling and control.

Keywords Energy storage · Battery modeling · Lead-acid batteries · State of charge (SOC) · State of health (SOH)

S. Blaifi (✉)
Research Laboratory of Electrical Engineering and Automatics LREA, University of Médéa, Médéa, Algeria
e-mail: sidali.blaifi@gmail.com

Department of Technology, Faculty of Sciences and Technology, University of Khemis-Miliana, Ain-Defla, Algeria

A. Mellit and M. Benghanem (eds.), *A Practical Guide for Advanced Methods in Solar Photovoltaic Systems*, Advanced Structured Materials 128, https://doi.org/10.1007/978-3-030-43473-1_8

8.1   Introduction

Energy supply for the remote areas that lacked of electrical network is one of the main concerns for the scientist researchers where certain autonomy is strongly needed. Unlike each system connected to the public network such as grid connected PV systems, the stand-alone PV systems are completely separated either in the form of isolated micro-grid or a simple stand-alone chain. These systems type have particularity in which they can continue to provide the energy in the case of lack of solar irradiation therefore, their name is usually related to the term energy storage. The storage in PV systems remains a major problem due to their unpredictable behavior. Several energy storage systems have been introduced in the practice however, the storage by battery is still widely used due to its low cost and its simple maintenance. However, the continuous changes of metrology conditions give a random change in the battery inputs (current and temperature) which make it complex in terms of modeling, control and real-state estimation. From there, in this chapter, a suitable battery modeling for PV systems is amply highlighted as well as the suitable state of charge estimation method. The adequate charging mode for these systems is also provided.

8.2   Electrochemical Storage

Electrochemical storage is the keep of electrical energy by transforming on electrochemical form to be provided to the load when needed. These storage systems are composed of three main parts namely, positive and negative electrodes isolated by electrolyte. Because of their low cost, the rechargeable batteries are commonly used for the storage of electrical energy. The battery can be formed of one or more cell collected in serial of parallel according to the desired sizing; each cell is composed of electrodes (anode and cathode) and electrolyte on liquid, solid or other form. The batteries are reversible systems which can behave as load in the charge process and as source in the discharge process. However, two main drawbacks are presented by the battery which somewhat limit its progress, the first is the complex electrical behavior that makes it difficult to be modeled, and the second is the capacity degradation wherein it can be unqualified after a certain lifetime. The recent researches are mainly focused on these two problems in order get a maximum lifetime and provide an accurate behavior prediction. From the material that the battery is composed, they exist several types such as Nickel–cadmium, nickel hydride, Lithium-ion and lead-acid battery...etc.

8.2.1 Nickel–Cadmium (NiCd)

This type of battery is formed of nickel hydroxide and metallic cadmium as electrodes, whereas the electrolyte is composed of the aqueous alkali solution. This type presents good performances in terms of the low maintenance and the high robustness. However, these batteries are not environmentally friendly from the fact that the cadmium and the nickel are toxic metals. Moreover, the memory effect can be appeared in this batteries type wherein their capacity decreases if they are usually recharged after partial discharge, as well as they present a high self-discharge current. The NiCd batteries are usually used for the starter systems and standby service.

8.2.2 Nickel–Metal Hydride (NiMH)

This batteries type is similar to the Nickel–cadmium, whereas it contains hydrogen-absorbing alloy in the negative electrode instead the toxic cadmium which makes it friendlier for the environment. These batteries present other advantages such as the low internal resistance and the high cycle durability. However, the high self-discharge current rate and the memory effect are still appearing in nickel–metal hydride-based batteries.

8.2.3 Lithium-Ion (Li-Ion)

For this batteries type, the electrodes are made of lithium metal oxide such as $LiCoO_2$ and $LiMO_2$ and graphitic carbon. The electrolyte is made of dissolved lithium salts such as $LiClO_4$. This batteries type is widely used in several electronic applications such as: smart phones, camera, laptop…etc. The lithium-ion batteries provide several advantages comparing to the other types, namely the high energy density, deepest discharge and lowest self-discharge current. The electrolyte in gel form makes this batteries type immune from leaking. However, the Li-ion-based batteries are very affected by the deeps discharges which accelerate considerably their capacity degradation. Also, these batteries are very affected by overcharging wherein they can be fully damaged after a high charging mode. As well as, an electronic pack is required for the monitoring of this batteries type that makes it relatively costly.

8.2.4 Lead-Acid Battery

Owing their low cost, these batteries technologies are the most used in the large-scale systems especially the PV systems. The positive and negative electrodes are made of lead dioxide (PbO$_2$) and of lead (Pb) respectively, whereas the electrolyte is composed of the sulfuric acid. The electrolyte takes part in the reaction of electrodes which is an important advantage of these designs. Lead-acid-based batteries present relatively a low self-discharge current and accept a deep discharge. As well as, the memory effect does not appear in this type of batteries. However, the lead-acid batteries have a limited lifetime comparing to other types wherein their capacity decreases with the cycling time. Moreover, they should be recycled as the lead is a toxic material.

Likely there is a consensus that the lithium-ion batteries present a better performances according to other types however, their high cost has slow down their progress especially in the large-scale applications. Consequently, the storage in photovoltaic stations is still practically done by using lead-acid battery.

8.3 Electrical Behavior of Lead-Acid Battery

In the charge and the discharge processes, the lead-acid battery passes through different areas which can affect significantly its lifetime. Wherein, for a nominal current (usually the current provided at 10 h), the battery crosses the charge, overcharge and saturation areas in the 16 h of charging mode, and passes through the discharge, over-discharge and exhaustion areas in the second 16 h of discharging mode. Figure 8.1 illustrates a charge/discharge voltage behavior of one element for nominal current. On the other hand, the working temperature is an important factor in the battery behavior. Where, its voltage can change by 5 mV if the working temperature changes by 1 °C. Moreover, the electrolyte temperature increases when the battery achieves the overcharge area. In addition, its lifetime can be strongly affected by the working temperature wherein the useful battery lifetime decreases by half if the working temperature increases by 10 °C [1–4].

Fig. 8.1 Working areas

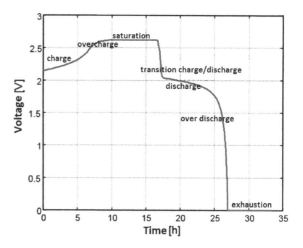

8.3.1 Charging and Discharging Areas

- **Charge area**

In this area, the battery stores the energy provided by an external source in electro-chemical form, where its voltage and its state of charge increase. This working area is handled as best in terms of battery health preservation. Indeed, no liberation of H_2O from electrolyte in this zone which involves the favorable operating of the battery. The maximum limit of this zone is when the voltage reaches the gassing level (and SOC <= 95%) [1–4].

- **Overcharge area**

The overcharge area is the zone limited between the gassing and the saturation levels wherein the battery starts to be saturated. In this area, the battery starts to liberate the H_2O from the electrolyte by gassing form which influences its charge density. The continuous operating in this zone affects considerably the battery health therefore, the charging source should be disconnected. Actually, the levels of these areas (gassing and saturation) change as function the operating temperature and the charging current which makes their estimation more complex [1–4].

- **Saturation area**

In this area, the battery is fully saturated and cannot accept more energy where the voltage stabilizes at its value and the current decreases until tends to zero. The continuous working in this area affects strongly its health therefore, the charging source should be isolated immediately [1–4].

- **Discharge area**

In the discharge process, the battery behaves as source to provide the energy to the load. Like the charging zone, this zone is handled as better in terms of battery health preservation. The limit of this area is when the voltage is higher than 90% of its nominal value whereas, the SOC indicator is more realistic to represent the provided energy in this zone relying on the discharge current value [1–4].

- **Over-discharge**

In this area, the battery is near to be empty and starts to discharge rapidly. The limit of this zone is when the voltage is lower than 90% of its nominal value. The continuous operating under this level affects considerably the battery health therefore, the discharging load should be disconnected [1–4].

- **Exhaustion**

This area starts when the voltage reaches 70% of its nominal value, wherein the battery is completely empty. It is very dangerous for the battery health, and it can be damaged if it is continually forced to operate in this area therefore, the discharging load should be isolated immediately [1–4].

• **Transition area**

Due to the battery internal resistance, in the transition from the charging to discharging mode and vice versa, a voltage difference is appearing. Wherein, the internal resistance of the battery can be determined by measuring this gap.

In the light of the aforementioned description, the accurate estimation of battery operating areas can play an important role in the working control which prolong and enjoy its lifetime [1–3].

8.4 Lead-Acid Battery Bank Modeling for PV Systems

The meaning of the battery bank term is the battery cells array associated in serial and in parallel to obtain the desired sizing on the voltage and the current. Among the different batteries types exist in the practice, the lead-acid battery remains the most used in PV systems due to its several advantages such as the low cost and the ease of manufacturing. However, the modeling of this batteries type is still a critical issue for this application wherein the battery is subjected to a continuous change of meteorology conditions. From there, many models of battery have been proposed in the literature based on different approaches such as: electrochemical, analytical and electrical. Whereas, the suitable approach for the PV systems application is the electrical.

8.4.1 Battery State of Charge (SOC) Estimation Methods

8.4.1.1 Definition of SOC Indicator

For any batteries type, the stored charge can be expressed by percentage indicator which called state of charge (SOC). This indicator gives an information about the quantity of amperes stored in the battery by electrochemical manner ($Q(t)$) according to its instantaneous capacity ($C(t)$). In the fact, the SOC is considered as the best indicator which provide the actual state of the battery in terms of operating areas.

$$\text{SOC}(t) = \frac{Q(t)}{C(t)} \tag{8.1}$$

Consequently, several methods of battery SOC estimation have been introduced in the literature which based on different approaches. It is worth noting that these proposed methods are applicable for any type of batteries especially for lead-acid technology [1–4].

Fig. 8.2 Stabilization voltage

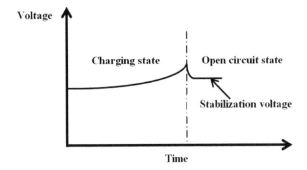

8.4.1.2 Classification of SOC Estimation Approaches

The main existing approaches of SOC estimation can be classified into three ways namely, direct measurement, mathematical methods and adaptive systems. On the other hand, each way may be combined with other to form hybrid estimation in the purpose to associate the advantages of each method to provide an optimal result.

Direct Measurement

This approach is based on the measure of physical characteristics of the battery such as the terminal voltage, open circuit voltage and the battery impedance. These methods give an accurate estimation of the battery SOC however, their first drawback is that they are not dynamic methods in which they involve the disconnection of the battery from the system. Then, these methods are recommended for the estimation of the initial battery SOC.

1. Open circuit voltage method

As expressed by Eq. (8.2), this method is based on a linear approximation between the SOC and the open circuit voltage V_{OC}. This approximation provides an accurate estimation of battery SOC however, as shown in Fig. 8.2, the open circuit voltage which gives the real information about the battery SOC is the stabilization voltage which is obtained when the battery is disconnected from the load and the source for a period longer than two hours, which consists of main drawback of this estimation method. Hence, this approach is not considered as dynamic method [5].

$$V_{OC}(t) = a_1 \times SOC(t) + a_0 \qquad (8.2)$$

where a_{0-1} are parameters identified by two point knowing the value of $V_{OC}(t)$ at $SOC(t) = 100\%$ and other point.

2. Terminal voltage method

The terminal voltage approach proposed by Anbuky et al. [6] is based on the discharge voltage drop caused by the battery internal impedance. In the fact that the battery electromotive force (EMF) varies proportionally with the terminal voltage and approximately linear with the battery SOC, the battery terminal voltage can be approximately linear with the battery SOC. This method has been validated for different discharge current and working temperature however, at the over-discharge area when discharge voltage starts to decrease rapidly, this method presents a significant estimation error [7].

3. Impedance method

This method consists of measuring of battery impedance by applying a low frequency AC signal to give knowledge of certain parameters which depend on the battery SOC. This approach requires a wide range of impedance experiments to identify and use of impedance parameters for the estimation of battery SOC. On the other hand, impedance spectroscopy method is employed which consists of measuring the battery impedances under wide range of AC frequencies for different charge and discharge currents. Then, the SOC is indirectly concluded by measuring several impedances at various SOC values [8].

Mathematical Methods

These methods are based on equations to estimate mathematically the battery SOC where the current is used as input of these equations. The most common mathematical methods have been employed are Coulomb counting method and modified Coulomb counting method. These approaches can be used for SOC estimation of any battery type however, their first drawback is the cumulative error during the sum calculation.

1. Coulomb counting method

As expressed in Eq. (8.3), this method consists of counting the sum of the incoming and the outgoing amperes pass from and to the battery added to the initial state of charge SOC_0, according to its instantaneous capacity $C(t)$ given in the empirical expression (8.4) provided by Shepherd [9]. Wherein, C_{tcoef}, A_{cap}, B_{cap} and α_c are empirical parameters, $C_{nominal}$ the battery the nominal capacity given by manufacturer, $I_{nominal}$ is the battery nominal current taken for 10 h ($n = 10$), T_{ref} is the reference temperature [1–4].

$$SOC(t) = SOC_0 + \frac{1}{C(t)} \int_0^t I(t)\partial t \tag{8.3}$$

$$C(t) = \frac{C_{nominal} \cdot C_{tcoef}}{1 + A_{cap}\left(\frac{|I(t)|}{I_{nominal}}\right)^{B_{cap}}} (1 + \alpha_c(T - T_{ref})) \tag{8.4}$$

$$I_{\text{nominal}} = \frac{C_{\text{nominal}}}{n} \tag{8.5}$$

These expressions can be modified by considering the degradation of capacity, self-discharge current and the battery state of health [3].

2. Modified Coulomb counting method

In the fact, this method is an improvement of the conventional Coulomb counting method by introducing a corrected current to enhance the estimation accuracy. As given in Eq. (8.6), it consists of defining a modified current $I_c(t)$ expressed by two degrees polynomial relationship.

$$I_C(t) = k_2 I(t)^2 + k_1 I(t) + k_0 \tag{8.6}$$

where k_{0-2} are parameters which can be identified based on experimental data.

Consequently, the battery SOC is expressed later by Eq. (8.7):

$$\text{SOC}(t) = \text{SOC}_0 + \frac{1}{C(t)} \int_0^t I_c(t) \partial t \tag{8.7}$$

The modified Coulomb counting method shows a better accuracy compared to the conventional method according to the experimental results [10].

Adaptive Systems

With the enhancement of artificial intelligence, several adaptive systems have been introduced in the literature based on non-parametrical estimation models such as artificial neural network (ANN), fuzzy neural network (FNN) and Kalman filter, where the nonlinear behavior of the battery SOC can be estimated. It is known that the non-parametrical model is extracted based on the real data therefore, their accuracy is related on the richness of this data in terms of scenarios [11]. On the other hand, the adaptive systems can be automatically adjusted to response to the complex changes due to many chemical factors [12].

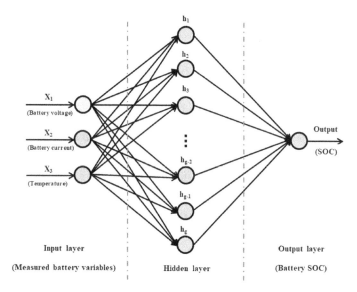

Fig. 8.3 Neural network architecture of battery SOC estimation

1. Artificial Neural network (ANN)

Due to its good ability to represent the nonlinear functions, the ANN is the most popular network type prevailed in the non-parametrical modeling. This approach is also applied to estimate the battery SOC where the model is built based on historical battery data to provide a relationship between a direct measured variables and the SOC. In which, the robustness of the obtained model relies strongly on the richness of the dataset [11].

As illustrated in Fig. 8.3, the architecture used to estimate the SOC is formed of three layers namely, input layer, hidden layer and output layer. The input layer contains three neurons for the measured battery variables namely, voltage, current and temperature, the hidden layer contains g neurons, and the output layer contains one neuron for the SOC [13].

According to the training approach, two types of ANN have been employed for SOC estimation namely, back propagation (BP) neural network and radial basis function (RBF) neural network. An accurate SOC estimation can be achieved however, the first drawback of these methods is the need of significant dataset [14].

2. Fuzzy logic method

The fuzzy logic methodology is also employed for the estimation of battery SOC which can give powerful means of modeling the complex systems especially the battery behavior. Salkind et al. [15] have proposed an experimental method to estimate the battery SOC with the use of fuzzy logic model to analyze the outcomes given by impedance spectroscopy and Coulomb counting-based methods.

The employment of fuzzy logic is also appeared for SOC estimation in the approach provided by Singh et al. [16] where AC impedance and voltage recovery measurements have been used as inputs of the fuzzy logic model. Singh et al. [17] have developed a fuzzy logic models for both available capacity and SOC estimation through the measure of the impedance at only three frequencies. A developed coulomb counting method is proposed by Malkhandi et al. [18], where the time-dependent variation has been compensated by using a learning system in which a fuzzy logic model has been used for the SOC estimation as component of this learning system.

3. Fuzzy neural network (FNN)

This approach provides a non-parametric model associated with fuzzy logic model to predict any complex and nonlinear behavior. The FNN has been introduced for several applications such as the FNN controller. This technique is also used by Lee et al. [19] for the battery SOC estimation modeling wherein a FNN with B-spline membership functions is used with an adjustment using genetic algorithm optimization algorithm.

4. Kalman filter (KF)

This estimator is based on the real-time measurement to predict a non-measurable variable. This approach has also been used for the battery SOC prediction. The KF-based method has been introduced by Xu et al. [20] for an accurate real-time estimation of the battery SOC.

In light of the previous description, it can appear that the coulomb counting method presents a good suitability with PV applications wherein it can be applied for a random variation of the current.

8.4.2 Battery Voltage Modeling

In this section, a battery modeling will be proposed by using CIEMAT model. This model has been developed for the PV applications which takes into account all battery characteristics such as state of charge (SOC), level of energy (LOE), instantaneous capacity (C) and the voltage output versus the current and the temperature [1–4].

8.4.2.1 Battery Cell Equivalent Circuit

Based on the simple representation illustrated in Fig. 8.4, the battery cell is represented by an open circuit voltage V_{oc} and internal resistance R_b resulting from the electrochemical energy stored in the battery cell and the different loss inside it, respectively.

The battery cell open circuit voltage varies as function the battery state of charge, where the stabilizing open circuit voltage can provide clear information about the

Fig. 8.4 Battery cell
equivalent circuit

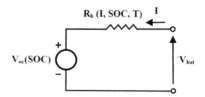

battery cell SOC, whereas the internal resistance varies as function charge/discharge
current, state of charge and working temperature. Also, the terminal battery cell
voltage given in Eq. (8.8) varies according to the working areas [1–4].

$$V_{bat} = V_{oc} \pm I \cdot R_b \tag{8.8}$$

where: the current I is positive in the charge process and negative for the discharge.

8.4.2.2 Temperature Effect

The temperature is an important factor in the battery cell behavior, where the working
temperature can affect strongly its lifetime and its output behavior such as voltage
and capacity. Typically, the battery cell voltage decreases 5 mV if the electrolyte
temperature increases by 1 °C. Moreover, the temperature of the battery cell electrolyte increases in the full charge areas such as overcharge and saturation. On the
other hand, according to rule of thumb, the useful battery cell lifetime increases by
half if the working temperature decreases by 10 °C. Hence, a cooling system should
be installed with the battery bank [1–4].

8.4.2.3 Battery Cell Capacity (C)

The capacity is a characteristic that consists of the quantity of current that the battery
cell is able to provide and absorb during a specific time. Equations (8.9) and (8.10)
express the evolution of this characteristic as function the charge/discharge current
and the temperature. Where, C_{tcoef}, A_{cap}, B_{cap}, α_c are empirical parameters which
will be identified, $C_{nominal}$ is the nominal capacity taken generally for 10 h (given by
manufacturer). T_{ref} is the temperature of reference taken 25 °C [1–4, 9].

$$C(t) = \frac{C_{nominal} \cdot C_{tcoef}}{1 + A_{cap}\left(\frac{|I(t)|}{I_{nominal}}\right)^{B_{cap}}}(1 + \alpha_c(T - T_{ref})) \tag{8.9}$$

$$I_{nominal} = \frac{C_{nominal}}{n} \tag{8.10}$$

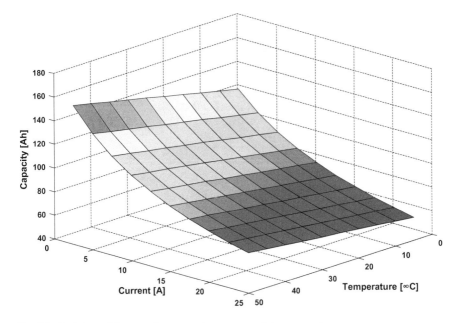

Fig. 8.5 Variation of the capacity versus the current and the temperature

Figure 8.5 describes the evolution of battery capacity versus the current and the temperature.

8.4.2.4 State of Charge (SOC) and Level of Energy (LOE)

The state of charge (SOC) provides information about the battery charge by percentage form. The method used in this model is the Coulomb counting method which provides dynamic estimation of this indicator.

As described in Eq. (8.11), the SOC is obtained by the sum of incoming and outgoing amperes from and to the battery cell divided on its instantaneous capacity. Besides, the initial state of charge SOC_0 should be also considered.

$$SOC(t) = SOC_0 + \frac{1}{C(t)} \int_0^t \eta_c(t) I(t) \partial t \tag{8.11}$$

The Coulomb efficiency η_c is a factor that describes the charge efficiency. This factor is considered in the charge process ($I > 0$) to express the energy loss inside the battery. Equation (8.12) shows the evolution of this factor where a_{cmt} and b_{cmt} are empirical parameters.

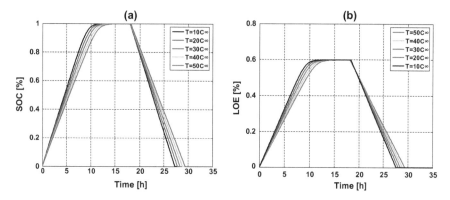

Fig. 8.6 SOC and LOE simulation for different temperatures

$$\eta_c = 1 - \exp\left(\frac{a_{cmt}}{\left(\frac{I}{I_{nominal}} + b_{cmt}\right)}(\text{SOC}(t) - 1)\right) \qquad (8.12)$$

The level of energy (LOE) describes as percentage the energy stored in the battery cell. It consists of the sum of the incoming and the outgoing amperes from and to the battery cell divided on the maximal capacity ($I = 0$, and $T = [-40, +40]$). Equations (8.13) and (8.14) show the variation of this indicator.

$$\text{LOE}(t) = \text{LOE}_0 + \frac{1}{C_n(t)}\int_0^t \eta_c(t)I(t)\partial t \qquad (8.13)$$

$$C_n = \max(C)\big|_{T=[-40,+40]}^{I=0} \qquad (8.14)$$

Figure 8.6a, b shows the temperature effect on the SOC and LOE, respectively. Figure 8.7 illustrates the evolution of the Coulomb efficiency during the charge and the discharge processes.

8.4.2.5 Internal Resistance

The internal resistance of the battery cell is a characteristic that describes the loss of energy inside it. The materiel constituting the electrolyte, the electrodes and connections presents a resistance that influences considerably on the battery cell behavior. This characteristic varies strongly as function the working temperature. Several methods exist to measure this resistance such as the current pulse, the period of rest and milli-ohmmeter-based method. Equation (8.15) shows an empirical expression of the battery cell internal resistance wherein P_1 to P_5 and a_T are empirical parameters [1–4].

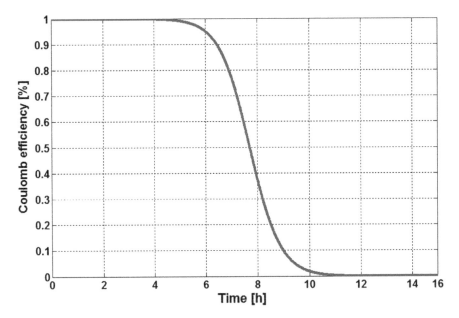

Fig. 8.7 Coulomb efficiency evolution during the charging process

$$R_b = \left(\frac{P_1}{1 + I(t)^{P_2}} + \frac{P_3}{(1 - SOC(t))^{P_4}} + P_5 \right)(1 - \alpha_T(T - T_{ref})) \qquad (8.15)$$

Surface graphs in Fig. 8.8a–c illustrate the variation of the battery cell internal resistance versus the current and the temperature for different states of charge.

8.4.2.6   Output Voltage

As is aforementioned, the battery passes through different areas during the charge and the discharge modes in which each area is expressed by empirical equation of voltage.

In the charge area ($I > 0$), the battery cell accepts the energy provided by the source and stores it in electrochemical form. The empirical expression (Eq. 8.16) describes the evolution of the battery cell voltage in this area. Where, P_{1c} to P_{5c}, V_{boc}, K_{boc}, and α_{rc} are empirical parameters to be identified.

$$V_c = (V_{boc} + K_{boc}SOC(t)) + \frac{I(t)}{C_{nominal}} \left(\frac{P_{1c}}{1 + I(t)^{P_{2c}}} + \frac{P_{3c}}{(1 - SOC(t))^{P_{4c}}} + P_{5c} \right)$$
$$(1 - \alpha_{rc}(T(t) - T_{ref})) \qquad (8.16)$$

The overcharge area starts when the battery cell reaches 95% of its charge. At this level, the battery cell begins to be saturated and liberates the hydrogen that is

Fig. 8.8 Variation of the
internal resistance versus the
current and the temperature.
a SOC = 80%, **b** SOC =
50%, **c** SOC = 30%

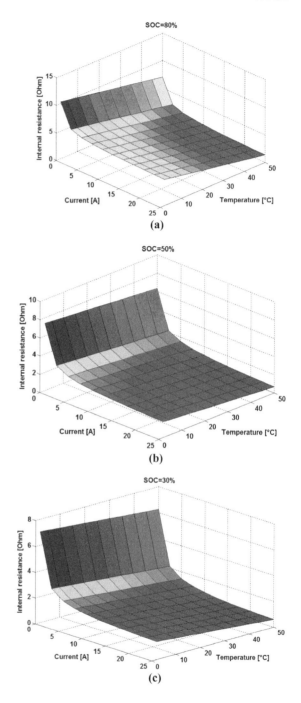

known as gassing voltage level. As is expressed in Eq. (8.17), the value of gassing level varies as function the charge current and the working temperature. Where, A_{gas}, B_{gas} and α_{gas} are empirical parameters to be identified.

$$V_g(t) = \left[A_{gas} + B_{gas} \ln\left(1 + \frac{I(t)}{C_{nominal}}\right) \right] (1 - \alpha_{gas}(T(t) - T_{ref})) \qquad (8.17)$$

If the battery cell continues to charge, it can achieve the saturation level, where it cannot accepts more energy. This area is very dangerous and affects strongly its lifetime, and therefore, the charging source should be disconnected immediately. Equation (8.18) shows the evolution of the voltage at this level. Where, A_{fonse}, B_{fonse} and α_{fc} are empirical parameters to be identified [1–4].

$$V_{ec}(t) = \left[A_{fonse} + B_{fonse} \ln\left(1 + \frac{I(t)}{C_{nominal}}\right) \right] (1 - \alpha_{fc}(T(t) - T_{ref})) \qquad (8.18)$$

Figure 8.9a, b illustrates, respectively, the variation of both gassing and saturation levels versus charging current and working temperature.

The voltage in the overcharge area is described in Eqs. (8.19)–(8.21) where SOC_{V_g} is the battery cell state of charge corresponds to the gassing voltage level, $A_{\tau cs}$, $B_{\tau cs}$ and $C_{\tau cs}$ are empirical parameters of the time constant to be identified.

$$V_{sc}(t) = V_g(t) + \left(V_{ec}(t) - V_g(t)\right) \cdot \exp\left[\frac{-(SOC(t)C(t) - SOC_{V_g}(t)C(t))}{I(t)\tau(t)}\right] \qquad (8.19)$$

$$\tau(t) = \frac{A_{\tau sc}}{1 + B_{\tau sc}\left(\dfrac{I(t)}{C_{nominal}}\right)^{C_{\tau sc}}} \qquad (8.20)$$

$$SOC_{V_g} = SOC\big|_{V_c = V_g} \qquad (8.21)$$

Equation (8.22) expresses the evolution of battery cell voltage in the discharge, over-discharge and exhaustion.

$$V_{dc} = (V_{bodc} - K_{bodc}(1 - SOC(t)))$$
$$- \frac{|I(t)|}{C_{nominal}}\left(\frac{P_{1dc}}{1 + |I(t)|^{P_{2dc}}} + \frac{P_{3dc}}{SOC(t)^{P_{4dc}}} + P_{5dc}\right)(1 - \alpha_{rdc}(T(t) - T_{ref})) \qquad (8.22)$$

In the transition from charge to the discharge or vice versa, the battery cell presents the "coupe de fouet" phenomenon which consists of a fast voltage variation where V_c at $I = 0 \neq V_{dc}$ at $I = 0$. To avoid the discontinuity at this point in the dynamic working, a linear function is introduced as is described in Eq. (8.23).

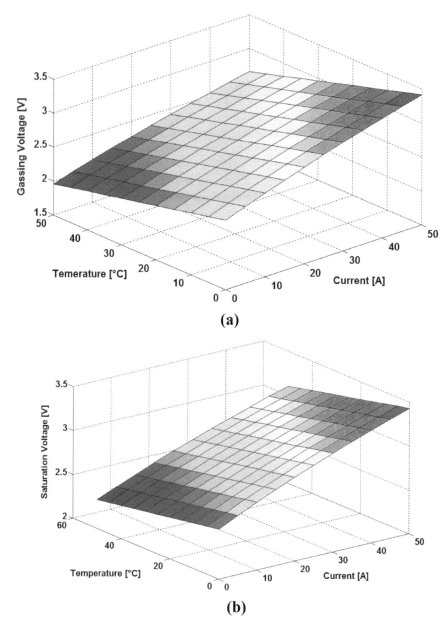

Fig. 8.9 Variation as function charging current and working temperature. **a** Gassing voltage, **b** saturation voltage

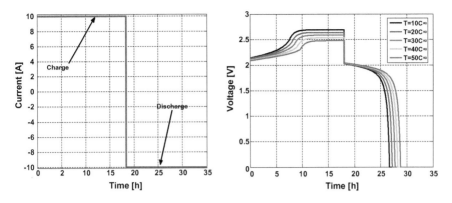

Fig. 8.10 Variation of battery cell voltage for the charge and the discharge

$$V_{cdc} = \frac{V_{c|I_\delta} - V_{dc|I_\delta}}{2I_\delta} I + \frac{V_{c|I_\delta} + V_{dc|I_\delta}}{2} \tag{8.23}$$

in which, I_δ is a small current that defines the transition band when $I_{bat} < |I_\delta|$.

Figure 8.10 shows the evolution of battery cell output voltage during the charge and the discharge for different working temperatures.

8.5 Model Parameters Extraction

In order to improve the accuracy of the aforementioned model, the empirical parameters can be identified using automatic parameters extraction process. For this end, suitable algorithm is introducednamely, genetic algorithm (GA). It consists of optimization process based on minimizing the error (cost function) between the real data and model output wherein 21 empirical parameters given in Table 8.1 are included. In which, the former parameters (introduced by Copetti et al. [4]) will be improved.

Table 8.1 Values of the CIEMAT model parameters

$C_{nominal} = 110$ Ah	$P_{1dc} = 4$ VAh	$P_{1c} = 6$ VAh	$A_{fonse} = 2.45$ V
$n = 10$ h	$P_{2dc} = 1.3$	$P_{2c} = 0.86$	$B_{fonse} = 2.011$ Vh
$C_{tcoef} = 1.67$	$P_{3dc} = 0.27$ Vh	$P_{3c} = 0.48$ VAh	$A_{\tau sc} = 17.3$ h
$\alpha_c = 0.005°$ C^{-1}	$P_{4dc} = 1.5$	$P_{4c} = 1.2$	$B_{\tau sc} = 852$ h
$\beta_c = 0°$ C$°$	$P_{5dc} = 0.02$ Vh	$P_{5c} = 0.036$ Vh	$C_{\tau sc} = 1.67$
$A_{cap} = 0.67$	$a_{cmd} = 20.73$	$A_{gas} = 2.24$ V	$\alpha_{rdc} = 0.007 °C^{-1}$
$B_{cap} = 0.9$	$b_{cmd} = 0.55$	$B_{gas} = 1.970$ Vh	$\alpha_{rc} = 0.025 °C^{-1}$
$V_{bodc} = 2.085$ V	$V_{boc} = 2$ V	$\alpha_{gas} = 0.002°$ C^{-1}	$I_\delta = 0.0001$ A
$K_{bodc} = 0.12$ V	$K_{boc} = 0.16$ V	$\alpha_{fc} = 0.002 °C^{-1}$	

Simulation of GA has been done using the algorithm parameters given in Table 8.2. The characteristics of both PV module and battery used in the experiment are summarized in Table 8.3. The parameters are regrouped into two groups. First group as given in Table 8.4 contains temperature independent parameters, and the second group as depicted in Table 8.5 contains temperature dependent parameters. Thus, parameters

Table 8.2 GA parameters taken in program

GA parameters	Value
Number of cycle	100
Population length	500
Crossover probability	0.7
Mutation probability	0.2

Table 8.3 PV modules and battery characteristics

PV module characteristics		Battery's nominal parameter	DC Lamps
$V_{oc} = 21.8$ V	$I_{mpp} = 7.67$ A	$V_{bat} = 12$ V	$V = 12$ V
$I_{sc} = 8.17$ A	$P_{mpp} = 135$ W	$C_{10} = 110$ Ah	$P = 11$ W
$V_{mpp} = 17.7$ V			

Table 8.4 Extracted Temperature independent parameters

Parameters	Extracted value	Parameters	Extracted value
SOC_0	0.7597	P_{1dc}	0.406 VAh
P_{1c}	7.234 Vah	P_{2dc}	3.041
P_{2c}	0.667	P_{3dc}	1.218 Vh
P_{3c}	0.078 VAh	P_{4dc}	0.7812
P_{4c}	0.492	P_{5dc}	0.484 Vh
P_{5c}	0.7421 Vh	V_{bodc}	2.148 V
V_{boc}	1.781 V	K_{bodc}	0.1270 V
K_{boc}	0.5313 V	B_{cap}	0.513
A_{cap}	3.351		

Table 8.5 Extracted temperature dependent parameters

Parameters	Extracted value
α_c	0.0081 °C^{-1}
C_{tcoef}	1.7871
α_{rdc}	0.0197 °C^{-1}
α_{rc}	0.43^{-1}

of the first group are identified for a fixed temperature, whereas parameters of the second group are identified separately.

After running GA for 100 cycles (Table 8.2 and Fig. 8.11a, b) using real measurements of 2 days taking one sample every 1 min. The following results are obtained wherein, Fig. 8.12a–e show the current provided by PV panel, the ambient temperature, both matching and mean error between the simulated model results and real measurements, respectively.

The obtained parameters by GA are shown in Table 8.4, for temperature independent parameters and Table 8.5 for dependent temperature parameters.

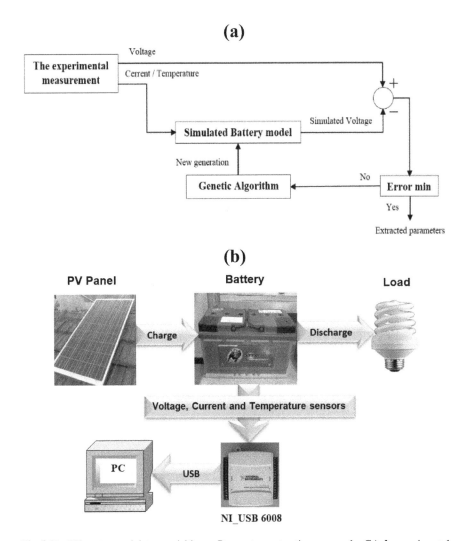

Fig. 8.11 PV system and data acquisition. **a** Parameters extraction process by GA, **b** experimental bench

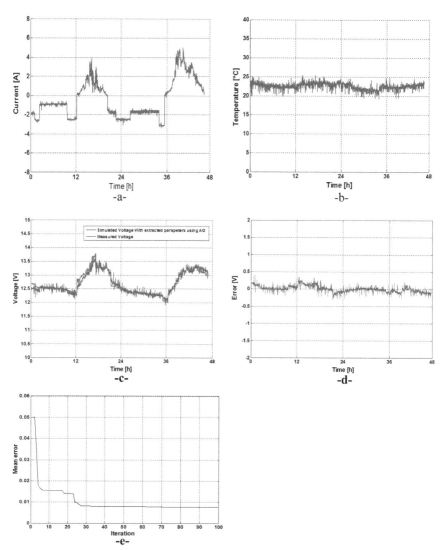

Fig. 8.12 Simulation and experimental results: **a** battery current measurement provided by PV module, **b** ambient battery temperature measurement, **c** measured and simulated battery voltage, **d** error between measured and simulated battery voltage, **e** the evolution of mean error

For the sake of more validation of the developed model in real conditions, real data measurements provided by PV modules (current and temperature) for different days are taken as battery input to compare the measured terminal battery voltage with that given by both the developed and the former CIEMAT models. A simulation was done with the lead-acid battery model taking the new extracted parameters after applying a GA and for a real current and temperature measure provided by the PV panel in different regimes (charge and discharge). We note that the energy provided by the PV system varies randomly due to the continuous change of radiation and temperature which gives a validation of the model in real condition. Figure 8.13(1–3) shows both the current provided by PV model that passes through battery and ambient temperature for three measurements in different days. Figure 8.14 shows the simulation provided by the improved battery model with extracted parameters using GA, compared to the experimental measurement and the simulation model with parameters given by CIEMAT in Table 8.1. It is obviously found that the improved model with new extracted parameters presents a good matching with the measurement

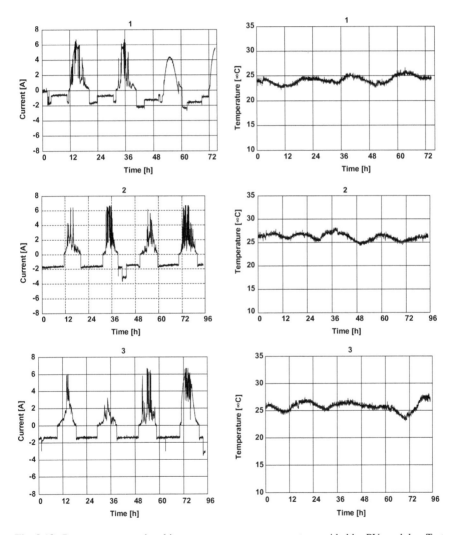

Fig. 8.13 Battery current and ambient temperature measurements provided by PV modules, Test 1: three days, Test 2: four days, Test 3: four days

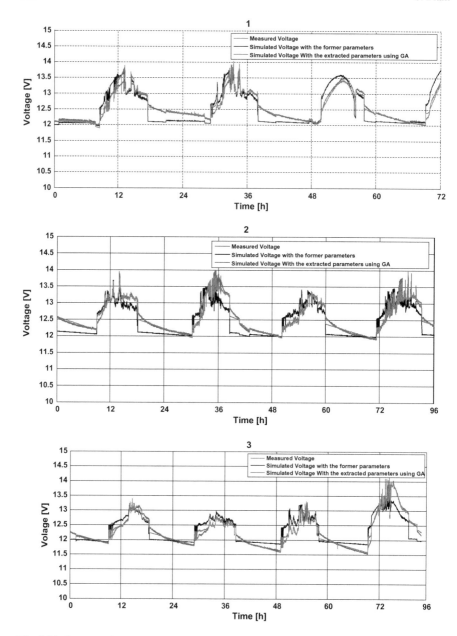

Fig. 8.14 Battery voltage measurements of three experiments and simulation results obtained by genetic algorithm extracted parameters and by former parameters

Table 8.6 Mean error simulated versus measured voltage

Simulation	1	2	3
Mean error with parameters given by CIEMAT (%)	1.54	1.84	1.52
Mean error with parameters extracted by GA (%)	0.50	0.48	0.46

comparing to the model with the former parameters and gives a lowest mean error as depicted in Table 8.6 wherein comparison between the mean errors obtained by simulation with parameters extracted using GA and simulation results from the former parameters is presented.

8.6   Conclusion

In this chapter, we have provided a highlight regarding the energy storage related to PV systems. The battery behavior has been amply highlighted beside the battery state of charge estimation methods. Moreover, a suitable modeling of the battery in PV systems has been provided as well as parameters extraction by using real outdoor measurement. Furthermore, charging control methods and controller's technologies have been explained. In light of the aforementioned description, it is obviously found that the modeling using CIEMAT model with parameters improvement gives promising results in terms of matching with real measurement. Also, PWM-based controller can be the adequate choice for the PV systems due to its advantage such as the possibility of MPPT integration using a single converter.

References

1. Blaifi, S., Moulahoum, S., Colak, I., Merrouche, W.: An enhanced dynamic model of battery using genetic algorithm suitable for photovoltaic applications. Appl. Energy **169**, 888–898 (2016)
2. Blaifi, S., Moulahoum, S., Colak, I., Merrouche, W.: Monitoring and enhanced dynamic modeling of battery by genetic algorithm using LabVIEW applied in photovoltaic system. Electr. Eng. **100**(2), 1021–1038 (2018)
3. Guasch, D., Silvestre, S.: Dynamic battery model for photovoltaic applications. Prog. Photovolt. Res. Appl. **11**, 193–206 (2003)
4. Copetti, J.B., Lorenzo, E., Chenlo, F.: A general battery model for PV system simulation. Prog. Photovolt. Res. Appl. **1**, 283–292 (1993)
5. Ng, K.S., Moo, C.S., Chen, Y.P., Hsieh, Y.C.: State-of-charge estimation for lead-acid batteries based on dynamic open circuit voltage. In: Proceedings of the 2nd IEEE International Power and Energy Conference (PECon'08), pp. 972–976, Johor Bahru, Malaysia, Dec 2008
6. Anbuky, A.H., Pascoe, P.E.: VRLA battery state-of charge estimation in telecommunication power systems. IEEE Trans. Ind. Electron. **47**(3), 565–573 (2000)
7. Sato, S., Kawamura, A.: A new estimation method of state of charge using terminal voltage and internal resistance for lead acid battery. In: Proceedings of the Power Conversion Conference, pp. 565–570, Osaka, Japan, Apr 2002

8. Rodrigues, S., Munichandraiah, N., Shukla, A.K.: A review of state-of-charge indication of batteries by means of A.C. impedance measurements. J. Power Sources **87**(1–2), 12–20 (2000)

9. Shepherd, C.M.: Design of primary and secondary cells. J. Electrochem. Soc. **112**, 657–664 (1965)

10. Ng, K.S., Moo, C.S., Chen, Y.P., Hsieh, Y.C.: Enhanced Coulomb counting method for estimating state-of-charge and state-of-health of lithium-ion batteries. Appl. Energy **86**(9), 1506–1511 (2009)

11. Blaifi, S., Moulahoum, S., Benkercha, R., Taghezouit, B., Saim, A.: M5P model tree based fast fuzzy maximum power point tracker. Sol. Energy **163**, 405–442 (2018)

12. Watrin, N., Blunier, B., Miraoui, A.: Review of adaptive systems for lithium batteries state-of-charge and state-of-health estimation. In: Proceedings of IEEE Transportation Electrification Conference and Expo, pp. 1–6, Dearborn, Mich, USA, June 2012

13. Weigert, T., Tian, Q., Lian, K.: State-of-charge prediction of batteries and battery-supercapacitor hybrids using artificial neural networks. J. Power Sources **196**(8), 4061–4066 (2011)

14. Linda, O., William, E.J., Huf, M., et al.: Intelligent neural network implementation for SOCI development of Li/CFx batteries. In: Proceedings of the 2nd International Symposium on Resilient Control Systems (ISRCS'09), pp. 57–62, Idaho Falls, Idaho, USA, Aug 2009

15. Salkind, A.J., Fennie, C., Singh, P., Atwater, T., Reisner, D.E.: Determination of state-of-charge and state-of-health of batteries by fuzzy logic methodology. J. Power Sources **80**(1–2), 293–300 (1999)

16. Singh, P., Vinjamuri, R., Wang, X., Reisner, D.: Design and implementation of a fuzzy logic-based state-of charge meter for Li-ion batteries used in portable defibrillators. J. Power Sources **162**(2), 829–836 (2006)

17. Singh, P., Fennie Jr., C., Reisner, D.: Fuzzy logic modeling of state-of charge and available capacity of nickel/metal hydride batteries. J. Power Sources **136**(2), 322–333 (2004)

18. Malkhandi, S.: Fuzzy logic-based learning system and estimation of state-of-charge of lead-acid battery. Eng. Appl. Artif. Intell. **19**(5), 479–485 (2006)

19. Lee, Y.S., Wang, W.Y., Kuo, T.Y.: Soft computing for battery state-of-charge (BSOC) estimation in battery string systems. IEEE Trans. Ind. Electron. **55**(1), 229–239 (2008)

20. Xu, L., Wang, J.P., Chen, Q.S.: Kalman fitering state of charge estimation for battery management system based on a stochastic fuzzy neural network battery model. Energy Convers. Manag. **53**(1), 33–39 (2012)

Chapter 9
The Flexibility in Power System with High Photovoltaic Penetration into Extra-High Voltage Level

S. Impram, S. Varbak Neşe, and B. Oral

Abstract The governments are adopting the use of renewable energy sources in electricity generation due to reasons such as the increasing climate change and energy supply security concerns and the rise of energy demand. Therefore, the share of variable renewable sources, especially solar energy, in total installed power capacity increases day by day. The power systems with the integration of solar energy sources transform. A power system to cope with high shares of variable solar generation needs to be flexible. In this chapter, the power system flexibility concept, the effect of variable renewable energy penetration, especially power plants on power systems flexibility, are examined. In addition, simulation studies are carried out for PV power systems penetration into extra-high voltage levels, necessary regulations for grid codes are determined, and solution methods are presented.

Keywords Power system stability · Photovoltaic power systems · Power grids · Power system stability · High-penetration photovoltaic

9.1 Power System Flexibility

Flexibility in power systems is the ability to provide supply–demand balance, maintain continuity in unexpected situations, and cope with uncertainty on the supply–demand side. The flexibility concept has changed in parallel with the historical development of power systems.

Initially, the flexibility in power systems was defined as the ability of the system generators to react to unexpected changes in load or system components [1]. Uncertainty and availability concepts are introduced on the supply side as a result of the increasing penetration levels of power generation from variable and hardly predictable sources such as wind energy (WE) and solar energy (SE), besides conventional generation systems. Hence, the flexibility of power systems has become a

S. Impram · S. Varbak Neşe (✉) · B. Oral
Marmara University, Istanbul, Turkey
e-mail: secil.varbak@marmara.edu.tr

concept that needs to be redefined. Therefore, new methods and management needs have emerged in order to provide flexibility in power systems where renewable energy (RE) penetration has increased.

Recently, the flexibility concept in power systems has taken place in the literature in a different way by organizations such as the International Energy Agency (IEA) and the North American Electric Reliability Corporation (NERC). Also, authors and groups have made their definitions in academic studies [2]. The IEA explains a power system as flexible, if it can, within economic boundaries, respond quickly to high fluctuations in supply and demand, ramping down a generation when demand decreases and upwards when it increases for scheduled and unpredictable events [1].

In the regenerated concept, besides the uncertainty on the demand side, there is also uncertainty on the supply side. In some references, flexibility is described as the ability of a power system to use its resources in order to respond to net load changes that are not served by variable generation [3–5]. According to a similar definition, flexibility is the ability of the power system to accommodate the net load changes by adjusting the in of adjustable loads or the output of generation units at certain time intervals [6]. In the study conducted by the Institute of Energy Economics at the University of Cologne on behalf of IEA [7], the flexibility concept defined as the capability to balance rapid changes due to RES generation and forecast errors. In another report of IEA [8], the capacity of the power system to modify generation and consumption in response to expected and unexpected variability is expressed as flexibility.

The supply and demand of the power systems are kept in balance, and the planning is done without the restriction of load changes. The necessary generation units are started up to provide balance by forecasting the load behavior [9]. From an operational perspective, flexibility is the potential of the generation to be deployed within a certain period [10]. Bucher et al. [11] define operational flexibility as the ability of the power system to damp the disturbances (such as generator trippings due to forecast errors or changes in the power injection) in order to protect the safe operating condition. Technically and economically, flexibility is the ability to cope with variability and uncertainty in both supply and demand while keeping reliability at a satisfactory level with a reasonable cost over various periods [12, 13].

9.1.1   Stability

A power system must be stable for proper and uninterrupted operation. The stability of the power system is defined as the ability to restore the operating balance after being subjected to a physical disturbance [14]. System stability is divided into three categories: rotor angle stability, voltage stability, and frequency stability.

Rotor angle stability is related to the ability of the synchronous generators that remain connected to the system after a failure. The occurred failure may cause loss of synchronization with other generators as distorting the rotor angle in some generators.

The transient stability, another form of rotor angle stability, includes system stability after major faults, such as loss of generation, line switching, or changes in load. The time range for this type of stability is usually 3–5 s and can be up to 10–20 s for large systems. The synchronization loss of a synchronous generator is influenced by many factors such as grid, generator and load parameters, grid structure, magnitude, and location of the fault [14, 15].

Frequency stability is the ability to keep system frequency constant after a failure that causes an imbalance between generation and consumption. The frequency stability problem is often related to insufficient equipment response, poor coordination between control and protection equipment, or insufficient generation reserves.

Voltage stability is the ability of the system to maintain a constant voltage level in all busbars under normal conditions and after a failure. The main reason for this instability is that the power system cannot meet the reactive power demand. Imbalances in voltage can cause rejection of loads from the system, faults in transmission lines, loss of synchronism in some generators, and voltage collapse [14].

One of the most important parameters for the synchronization of power systems is the system inertia. As the power system, inertial response decreases, its sensitivity to frequency deviations increases [15]. RE plants do not contribute to the system inertia as they are connected to the grid utilizing power electronic components and electrically separated from the grid. On the other hand, SE systems do not contribute to inertia in any way due to their physical structure and reduce the system inertial response [16].

9.1.2   Flexibility Parameters

Flexibility in conventional power systems is ensured by providing reserves and generation planning. Demand in such systems is highly predictable; hourly changes can be balanced with regulation and load tracking reserves. A reserve is also kept for unexpected generation outages or transmission line failures.

In modern power systems, besides the conventional generation units, there is variable generation (VG) with shares depending on the level of penetration. Conventional generation plants track changes in net load. The net load is the load that cannot be met by a variable sourced generation. Depending on the penetration levels of the variable RESs, the net load may have different properties than the normal load. Increased variability and different ramping models are examples of these features. Such features produce more flexibility need [17, 18]. Generation flexibility in power systems is based on the three basic parameters, absolute power output range (MW), ramping rate (MW/min), and energy level continuity (MWh), as shown in Fig. 9.1 [6, 17, 19, 20].

- The absolute power output range is the difference between the installed power of a plant and the minimum power that it can operate in a stable condition. The largeness of this difference can provide flexibility in wider system conditions.

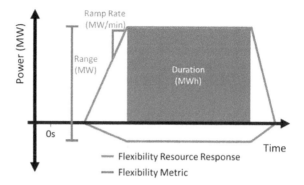

Fig. 9.1 Three dimensions of flexibility: power range, ramp rate, and continuity [17]

- The ramp rate shows how quickly the power plant can change its output power. The sources that have a high ramp rate are more flexible.
- The energy level continuity shows the duration of a certain power output level that the source can provide. The sources with a long duration increase flexibility by meeting demand under long-term faults or outages.

In power systems, the flexibility of all components is preferred. These components are classified into four categories [21].

- In a flexible generation, power plants should have a high ramp rate and be able to be operated at a low output power.
- Flexible transmission lines should be able to access various balancing sources to provide adequate capacity in case of congestion. Interconnections with neighbor power systems and the use of smart grid technologies improve system flexibility.
- The instant and direct control of demand response, storage, and flexible distributed generation should be achieved using smart grids for demand-side sources.
- Several practices can increase the flexibility of the existing system. Some of them are making decisions in near real-time, use of more sophisticated weather forecasts, and collaborating better with neighboring systems.

However, the flexibility indicators in a power system are changes in load, weather forecast errors, outages of generation plants or transmission line equipment, and generation from variable RE [22]. Also, in the report of the NERC's Integration of Variable Generation Task Force (IVGTF), three key features have been identified that need to be considered when assessing the flexibility requirements in the power system [23]. These are the magnitude of net load changes, the time interval over which these changes occur, and the frequency of ramping events. Thus, it is aimed to be able to distinguish periodic ramps and unpredictable ramps and to provide the required flexibility for balancing a net load ramp in a certain time interval by using the available resources properly [5].

9.1.3   Renewable Penetration Effects on Power System

The need for sudden, high ramping, and frequent start-ups caused by the intermittent and variable nature of the electricity generation from RES, coupled with the net load changes, causes difficulties for conventional generation units [12, 13].

The variability concept is considered differently at every stage of planning and operation. While the net load changes do not play an important role in long-term resource planning, the diurnal cycle of weather conditions is a major factor in the day ahead operation plan [13]. In the time interval (ms) that can be defined as a very short term, some control systems are required due to the instantaneous changes in the RE generation. These are control systems for low-voltage ride through (LVRT) [24] active and reactive power, voltage, and ramp rate [25, 26].

The operational flexibility type depends on the time scale. While more frequency control and reserves are needed in the seconds to minutes time interval, there is a need of increased ramping capability for minutes to hours and planning flexibility for hours to a day ahead time interval [13]. The flexibility time interval and variable generation effects from the system planning perspective are shown in Fig. 9.2 [13].

In the case of a large proportion or all of the demand is supplied by RES generation, the base load plants must reduce or completely stop their generation. However, these plants should be re-dispatched to meet the demand with a decrease in RES generation. This case constitutes a big problem since the start-up times of coal or nuclear power plants are too long [26]. As a result of exposure of these plants to excessive cycling, more wear and tear, metal fatigue, corrosion, and erosion are expected in the medium term, especially in the components where there are high temperatures and pressure [27–29]. This condition leads to an increase in operation, maintenance, and fuel costs and a decrease in the life of the plants. Also, the plant outages for maintenance purposes are more frequent [26, 27, 30]. In the long term, it will be a transition to low-carbon solutions in base load plant technologies with carbon law and limitations. It is also important that these are more flexible than existing technologies [26]. VRE penetration levels of countries at the end of 2018 that are an indicator of rapid transition are given in Fig. 9.3 [31]. Existing power system flexibility makes it easy to integrate PV electricity initially, but integrating larger shares will require

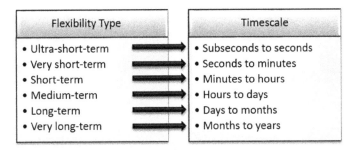

Fig. 9.2 Effects of variable generation on the flexibility timeline

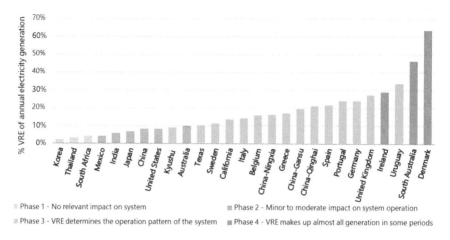

Fig. 9.3 VRE shares in total electricity generation by countries, 2018

greater flexibility. Higher power system flexibility enables more PVPP integration. IEA classified VRE integration according to shares is shown in Fig. 9.4 [31]. In phases 1 and 2, VRE integration management is easy and made by existing power system flexibility. In addition, policy-makers attention contributes to integration. In phases 3 and 4, VRE penetration is large. Hence, in these phases, flexibility can be required and achieved by implementing demand-side response (DSR), expanding grid interconnections, raising hydropower capacity, using CSP and other thermal plants flexibly, and eventually expanding storage from pumped storage hydropower and batteries. In phases 5 and 6, weather conditions and the time difference between demand and generation are the primary impediments for integration [31].

Innovations are needed in the planning and operation of transmission networks. The main purpose of the existing transmission lines is to transmit the energy from the

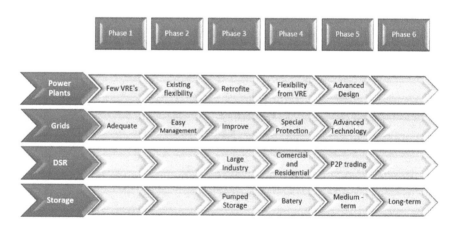

Fig. 9.4 Six phases of VRE integration and the four pillars of flexibility

regional generation units to the load centers. However, the distance and voltage levels are increasing with the installation of RES plants far away from the load centers at the endpoints of the network.

On the other hand, the RES generation, which is distributed over a broader area, decreases the variability of total generation, and this advantage can be utilized with correct planning [32]. The increase in penetration levels requires finding the optimal network topology. The network topology has a serious impact on transmission line losses and overall system performance in the case of a fault [33].

9.2 The Growing PV System Penetration Effects on Power System

Since PV systems started to become widespread, they were used as small and distributed systems, and they were connected to the network at the distribution level until 2009. However, a PV system was connected to the transmission system for the first time in the USA in 2009 [34]. In addition, interest in the connection of PVs to high and extra-high voltage lines has increased, and in countries such as the USA, China, India, Germany, and Spain, the installation of PVs connected to high voltage lines has been realized [35, 36].

9.2.1 Impact of PV Systems on Power System Stability and Flexibility

A study on PVs voltage and reactive power responses was performed by the California Independent System Operator (CAISO) [37] for various connection types. It has been indicated that overvoltage problems are inevitable due to the high share of PV connected to the sub-transmission network in the study. It was also pointed out that in a system with PV, Static Var Compensators (SVCs) caused higher transient overvoltages. The main reason for this was stated as the injection of reactive power for several cycles into the system by the SVCs due to their low operating speed after the clearance of the fault. According to the steady-state analysis that was performed for PV penetration by Eftekharnejad et al. [38], the most affected system parameters are voltage magnitudes. Overvoltages occurred in the transmission line busbars, especially at 20% and some other penetration levels. In a system with high PV penetration after transient events, larger voltage drops were found after a fault. Also, the disconnection of a large part of the rooftop PV systems resulted in increased voltage fluctuations and damping times as the penetration levels increased. In another study, the effects of centralized and distributed PV systems on the steady-state voltage stability of the Ontario power system were investigated. Various penetration levels with an installed power of up to 2000 MW have been examined. The results showed that the distributed PV could greatly improve voltage stability compared to the central systems [39].

In studies performed for transient stability, the effect of penetration level was firstly investigated at a 5–30% range. The results show that improving the system stability for penetration levels above 10% of PV depends on the fault ride through (FRT) capabilities of these power plants [37]. Eftekharnejad et al. [38] investigated the effects of rooftop and large-scale PV systems penetration on a large interconnected power system. For this purpose, PV penetration levels of up to 50% were examined by reducing the share of conventional generation. Analyzes have shown that high PV penetration levels have both positive and negative effects on the system transient stability. In addition, PV penetration levels, system topology, type, and location of the fault are key factors for the nature of the effect (positive or negative).

Furthermore, the disconnection of a large proportion of rooftop PV causes deviations in the rotor angles of nearby synchronous generators (SGs) and voltage fluctuations. In another study, the effects of large-scale and distributed PV integration on the transient stability of the Ontario power system have been examined. Critical clearing time (CCT) indication was used to evaluate the system's dynamic stability performance. A three-phase short-circuit fault for 80 ms in the 500 kV transmission line in the Toronto area has been tested. The results show that central PV power plants with voltage and reactive power control do not change the system's dynamic stability. On the other hand, an increase in distributed PV penetration levels improves transient performance [39].

Studies have also been conducted on the effects of PV penetration on small-signal stability. Liu et al. [40] examined the effects of the location and penetration level of PV generation on the two-area power system. The results show that the effect of the high PV penetration level is positive or negative, depending on the state of the SGs that are displaced. In another research, a 3-SG 9-bus test system was used, and the system was modified with the real-time data of the Indian network. The effects of variables such as solar irradiation, temperature, load, and configuration have been investigated. An increase in rotor modes has been observed in the integration of PV into the network and in the same way with increasing solar irradiation. Furthermore, the damping of the modes has also increased with increasing load, while there is a decrease with increasing generation [41]. Du et al. [42] used a single-machine infinite bus power system in their study. Analyzes show that PV generation affects small signal stability by interacting with conventional generation due to the lack of rotating components. However, there are not any additional oscillation modes added to the system. This effect varies depending on the operating conditions of the system since the contribution of the damping torque of the PV power plant can be positive or negative. After a certain critical operating condition, the effect of PV generation on the system small signal stability becomes negative. Eftekharnejad et al. [43], in another study, used large-scale PVs and rooftop PVs, which are aggregated at the voltage level of 69 kV. According to the results, there is a significant reduction in damping ratio as large conventional generators are displaced, while penetration level increases from 30 to 40%. The increase in penetration level causes a decrease in the system inertia resulting in a decrease of critical modes damping of the system.

In the studies of PV penetration in terms of frequency stability, Alquthami et al. [44] have assessed penetration levels of 5, 10, and 20% while keeping SGs in the system. Simulations show that the system frequency stability is adversely affected at 20% penetration level. In another study carried out in a two-area power system, the authors used a real-time simulation model of 4 × 50 MW PV system. In this study, automatic generation control (AGC) was applied to allow the maximum penetration level by adjusting the output power of the generators since SGs are not disconnected. Increased penetration level in one region of the system led to positive effects in both regions, such as faster damping of frequency oscillations and lower magnitude (overshoot) of oscillations [45].

In a system where a significant amount of SGs is disconnected, the control systems and their coordination are affected as well. Besides, fault state characteristics exhibit different features. The fault current of SGs is 5–10 times the nominal current, while it is roughly two times for inverter-based systems and decreases with time. This case sometimes prevents the protective relays from detecting the fault conditions in inverter-based systems [46, 47]. On the other hand, the inverters can avoid the thermal overloading of the network components by the rapid response to network imbalances. Another advantage of inverters is that their fault currents can be programmed [46].

9.2.2 Modeling and Simulation

In this section, simulations are conducted in order to analyze the impacts mentioned above transmission connected PV systems on power system stability. DIgSILENT® PowerFactory is used as a software tool for modeling of a PV system, power system, and application of transient stability analysis. This calculation software is a computer-aided engineering tool for the analysis of transmission networks, distribution networks, and industrial power systems. It is an advanced, integrated, and interactive software package to use in planning and operating optimization, which are the main objectives of power and control systems. Many transmission and distribution system operators, research centers, and universities around the world use this simulation software.

In many regions, generation plants from RESs are integrated into networks, leading to significant structural and topological changes in power systems. The enlargement and increasing complexity of power systems require various engineering and analysis studies in terms of industrial and commercial applications during the planning and operation. Inefficient planning and operating of a network lead to cost loss.

PowerFactory software offers a wide range of power systems analysis. Its functions include load flow analysis, optimal load flow, short circuit analysis, network congestion analysis, power quality and harmonic analysis, network reduction, arc flash analysis, techno-economic analysis, small-signal analysis, and system stability analysis. It also has a big library that includes components such as line conductors/cables, loads, generators, transformers, busbars, relays, and breakers. In this way, it is possible to create a model with components that are accepted or commonly used according to international standards.

In order to observe the response of the transmission system to increased PV system penetration, dynamic simulation results of the model should be examined. The transient simulation functions in the software analyze the dynamic response of large and small-scale systems in the time domain. Thus, it enables detailed modeling of complex and large transmission systems considering electrical and mechanical parameters [48].

9.2.2.1   The Test System

The test system, which is used for the simulations, is a built-in transmission system model of DIgSILENT. Various shares of VRE are simulated, adding additional PV systems and an aggregated dynamic model, which represents a large-scale PV system, is used in the simulation. As the output power of PV systems varies with solar radiation, power system behavior during these fluctuations is evaluated.

In order to observe the effect of increased PV systems penetration into the extra-high voltage level on the power system, simulations were performed in the transmission line model of DIgSILENT (Figs. 9.5, 9.6, 9.7, 9.8 and 9.9).

- This system consists of four regions, 13 substations, and 25 busbars.

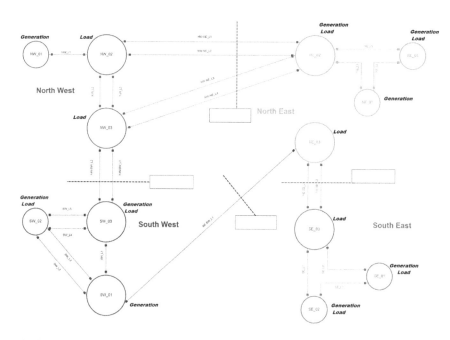

Fig. 9.5 General view of generation and load busbars in the regions

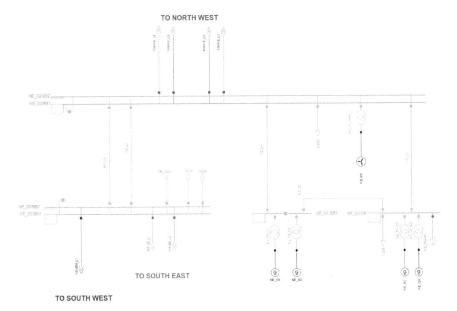

Fig. 9.6 Single line diagram of northeast region

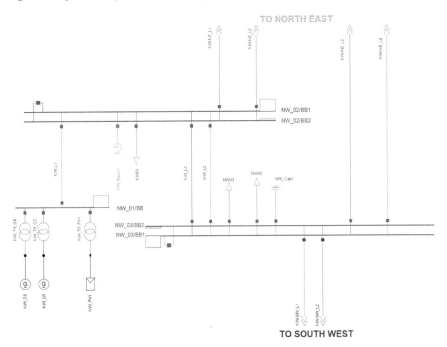

Fig. 9.7 Single line diagram of northwest region

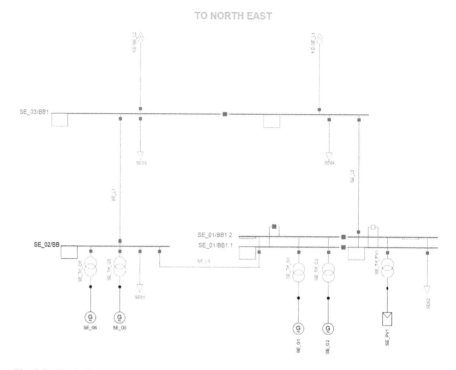

Fig. 9.8 Single line diagram of southeast region

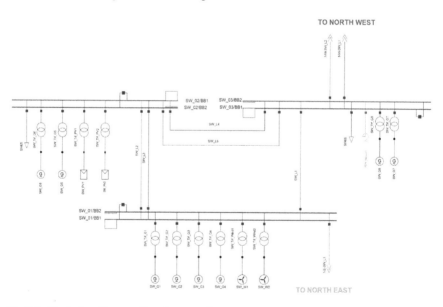

Fig. 9.9 Single line diagram of southwest region

- Energy transmission is carried out at 400 kV level through 25 lines.
- There are 18 SGs, 3 WE, and 4 PV systems as RES for electricity generation.
- It reflects the diversity of thermal power plants in a real power system, with the fact that SGs fuels are nuclear, coal, natural gas, and liquid fuels.
- The total installed capacity of the system is 9918 MW with 8568 MW thermal, 750 MW wind, and 600 MW solar.
- According to these values, the ratio of RE to the installed power is 13.6%. On the other hand, the total demand is 6050 MW [48].

Each SG in the system has automatic voltage regulator (AVR), power system stabilizer (PSS), and turbine governor. The turbine governor helps regulate the system frequency by regulating turbine speed and hence power. It controls the amount of steam going into the turbine and the amount of gas going into the combustion chamber in the gas turbines, depending on the power and speed error [49]. The main function of PSS is to damp the generator rotor oscillations in the range of 0.1–2.5 Hz at the system, which are called electromechanical oscillations by Kundur [50]. The output signal of the PSS, a feedback controller, is added to the summing point at the input of the AVR [51]. Finally, the task of the AVR is to control the voltage and reactive power change by acting on the excitation system of the generator. In cases such as fault and overload in the system, the generator excitation voltage is regulated by AVR to ensure that the end voltage is kept within limits allowed by the system stability [52].

9.2.2.2 Case Studies

The existing RESs in the system are predominantly in the southwest region. Assuming that the resources in the region are more efficient, this region has been selected to add PVs. It is also important to note that the existing RES and the added PVs are directly connected to the transmission system at 400 kV voltage level via transformers. Considering that RE will replace conventional systems, SG that is equivalent to installed power of added PVs has been disabled. In some cases, disabled conventional is not equal to the added power due to single units and high power of generators.

It is assumed that the existing PVs in the model does not generate during the simulation. The power output of each RES is 115 MW. The solar systems existing in the DIgSILENT PowerFactory software were used to model the PVs. The solar panel model that has 500 kW power and 0.9 power factor was used for PVs. The total power of power plants can be increased by entering the number of parallel inverters in the solar panel. In the solar panel model, there are solar radiation and temperature inputs, active and reactive power control, and control of active power drop for high frequency. The output voltage of the solar panel is 0.4 kV. In order to connect to the transmission system without using two set up transformers, the output voltage of the PVs is assumed as 16.5 kV. For this reason, 16.5/400 kV transformers are used for the connection of PVs in the DIgSILENT transmission system model.

The reason for choosing the model that allows easy operation and high penetration levels is to observe the initial stage of challenges at VRE penetration into extra-high voltage level for the interconnected system. The range of 5–10% penetration levels is considered as low level. So, in this part, the higher penetration levels are observed. The required PVs were added to reach 15, 20, and 30% penetration rates that will be investigated. In a power system, protective devices take actions after 11–30 min time range for the necessary operations in frequency imbalances. Therefore, protective devices have been deactivated to examine the power system response clearly. Also, the outputs of the added PV power plants were changed every 15 min to simulate solar energy variability. The total simulation time is 1 h and 15 min, and added plants are running at full power for the first 15 min. The plants operate at 75, 50, and 25% of nominal power for each following 15-min periods, respectively. At the 60th minute of the simulation, the power output of the added plants is zero.

Under normal operation conditions, frequency is stable at 49.999 Hz, and the voltage value is stable at 0.999 pu for busbar SW_03-BB1 that is in the PVPP region. These results show that the system is stable under normal operation conditions, and the model reflects a real power system operation. Load flow analysis under normal operation condition is shown in Fig. 9.10. As in the load flow analysis figure, there is no overloaded transmission line in the network.

In case 1 (15% solar energy): To reach 15% solar energy, the installed PV capacity of the system needs to be increased by 150 MW in total. Therefore, two 75 MW solar

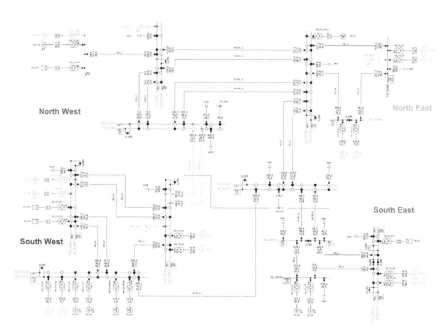

Fig. 9.10 Load flow analysis of the model under normal operation conditions

power plants were included in the system. Besides, a 450 MVA natural gas-fired thermal plant was deactivated.

In case 2 (20% solar energy): For the next level, 20% to be tested, the addition of 650 MW PV to the system is required. 2X250 MW and 2X75 MW solar power plants were added to busbars in the southwest region. At the same time, 592 MVA natural gas-fired SG was also deactivated.

In case 3 (30% solar energy): Finally, the 30% PV share was tested. In order to achieve this ratio, 1650 MW PV is added to the system. 6X250MVA and 2X75MVA PV were equally distributed to busbars in the southeast. In addition, two coal plants, each with a power of 592 MVA and a natural gas-fired power plant with a power of 450 MVA, were deactivated. The total power of the disabled SGs is 1634 MVA.

For power systems, the frequency band is normally accepted as 49–51 Hz (±2%) and the voltage band as 0.9–1.1 pu (±10%) for normal conditions in grid codes. For all examined penetration levels in the power system model, voltage values of busbars in all four regions and system frequency values are given in Tables 9.1 and 9.2. When the table is examined, it is seen that busbar voltages for all penetration levels are in the range of 0.9–1.1 pu limit values. One important point to consider when evaluating this situation is that supply 9918 MW and demand 6050 MW in the model under normal operating conditions. The power system model operates under conditions where supply can easily meet demand. In addition, at all penetration levels, system

Table 9.1 Busbar voltages in southwest, southeast, and northwest regions in the power system model for all case studies

PVPPs output power (%)	RE penetration rate (%)	SW_03-BB1		SE_01-BB1.2		NW_01-BB	
		Time (s)	Voltage (pu)	Time (s)	Voltage (pu)	Time (s)	Voltage (pu)
100	15	453.850	1.001	457.180	0.999	440.450	1.012
100	20	408.220	0.999	409.637	0.999	373.317	1.012
100	30	283.900	0.989	437.580	0.997	342.580	1.007
75	15	1351.322	1.001	1345.982	0.998	1316.472	1.012
75	20	1312.202	0.998	1289.722	0.998	1259.052	1.012
75	30	1169.632	0.991	1326.132	0.996	1228.332	1.008
50	15	2251.392	1.001	2251.262	0.998	2221.892	1.012
50	20	2224.652	0.998	2206.292	0.997	2147.582	1.011
50	30	2116.852	0.991	2228.612	0.994	2139.292	1.007
25	15	3167.522	1.001	3111.352	0.998	3154.192	1.011
25	20	3040.352	0.998	3055.802	0.996	3033.252	1.011
25	30	3047.412	0.990	3064.073	0.991	2997.132	1.006
0	15	4046.232	1.000	4059.592	0.998	4064.842	1.011
0	20	3969.922	0.997	3949.932	0.995	3921.902	1.010
0	30	3949.842	0.986	3986.132	0.988	3958.122	1.002

Table 9.2 Voltage of northeast busbar and frequency in the power system model for all case studies

PVPPs output power (%)	RE penetration rate (%)	NE_04-BB		SW_03-BB1	
		Time (s)	Voltage (pu)	Time (s)	Frequency (Hz)
100	15	448.490	1.002	419.090	49.996
100	20	423.607	1.002	362.600	49.999
100	30	364.930	0.999	306.250	49.998
75	15	1345.972	1.002	1351.242	49.982
75	20	1259.052	1.002	1312.192	49.943
75	30	1273.072	0.999	1289.812	49.834
50	15	2251.402	1.002	2251.352	49.966
50	20	2147.502	1.001	2241.812	49.873
50	30	2153.292	0.999	2139.222	49.528
25	15	3119.502	1.001	3127.422	49.955
25	20	3086.842	1.001	3034.562	49.771
25	30	3212.242	0.997	3041.782	49.233
0	15	4065.012	1.001	4064.902	49.941
0	20	4025.382	1.000	4029.872	49.665
0	30	3958.292	0.995	3921.922	48.888

frequency is in the range of 49–51 Hz normal condition limit values even while VRE-based generation is at the lowest generation level. Even in lack of VRE-based generation at 15 and 20% levels, frequency is within the normal condition limit values range. In lack of VRE-based generation at 30% penetration level, frequency is out of normal condition limits with 48.888 Hz.

In addition, smaller power systems such as the used model require a narrower frequency band for disconnection, for example, Malta and Canary Islands. For the most of a year, the normal frequency range on Malta is in the range of 49.5–50.5 Hz. In the Canary Islands, frequency is in range of 49.85–50.25 Hz [53]. In case lack of VRE-based generation at 20% penetration level, normal operation limit values are exceeded with 49.233 Hz. In addition, the voltage response of all busbars and frequency response of the system for the highest penetration level of 30% are presented in Figs. 9.11, 9.12, 9.13, 9.14 and 9.15.

9.3 Grid Codes and Operation

The regulations required for the reliable, stable, and low-cost operation of a power system are called grid codes. These codes cover the obligations of the transmission system users and other users who are connected to the distribution system but affect the transmission system. They also consist of the plant design and operation rules that

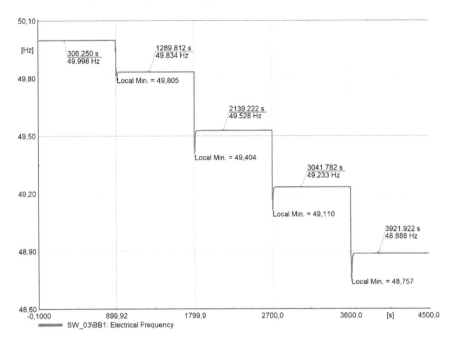

Fig. 9.11   Frequency response of SW_03-BB1 busbar in the southwest region

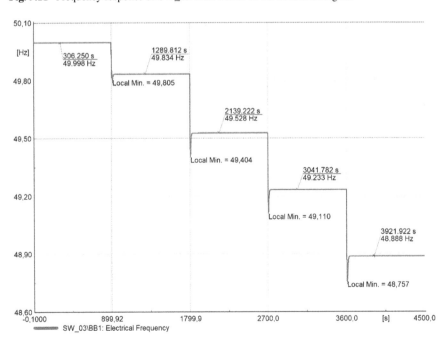

Fig. 9.12   Voltage response of SW_03-BB1 busbar in the southwest region

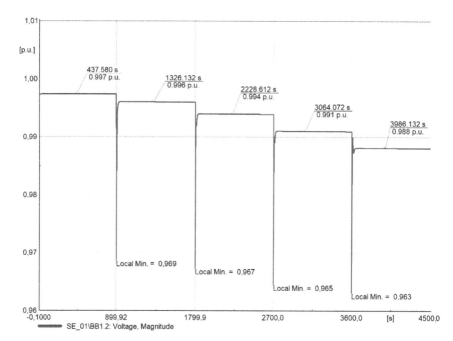

Fig. 9.13 Voltage response of SE_01-BB1.2 busbar

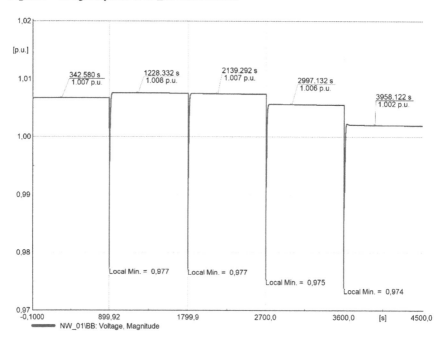

Fig. 9.14 Voltage response of NW_01-BB busbar

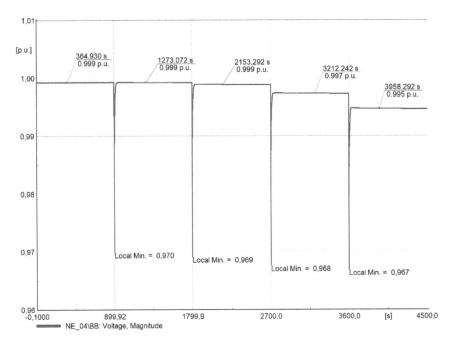

Fig. 9.15 Voltage response of NE_04-BB busbar

they must comply with, the issues to be observed for operating system considering the conditions related to transmission system planning and system security [54].

These codes have traditionally been developed for the connection of synchronous generator-based power plants to the grid. On the other hand, although the share of RES for electricity generation at first was very low compared to conventional power plants, it has increased recently. Therefore, to operate the power system stable, there was a need to prepare codes for RE generation plants.

Currently, the standards required for the integration of RESs are national. As taking into account the past experiences and practices of other countries, countries update grid codes when necessary. Grid codes in different countries are similar since the basic objectives are the same.

Although they are generally similar, grid codes may vary in content or the requested values for various reasons. How the network is managed throughout the country, the development level of generation from RE, power system characteristics, and the different operators can be shown as examples of these reasons [55].

According to technical requirements, voltage tolerance is normally within ±10% of the nominal value. On the other hand, the frequency tolerance is much less and in the range of ±2% in large interconnected systems. Apart from the given tolerances, the plants must remain connected to the system for a certain period of time depending on the fault magnitude. These determined time intervals allow duration for the system operator to take action against faults.

The technical requirements that should be added to the grid codes according to the penetration level of the variable RESs determined by the International Renewable Energy Agency (IRENA) are listed in Fig. 9.16 [53].

Since the wind power plant (WPP) is a more common and mature technology, the focus is on the connection of them to the grid. Only countries like Germany, Denmark, Spain, and Italy are looking for specific technical requirements for PVs.

The development of grid codes should be carried out in coordination with energy policies. Otherwise, problems may arise in terms of operating the system. In Germany, for example, in the period of low PV shares, it was mandatory for the power plants connected to the low voltage (LV) network to be disconnected when the system frequency was above 50.2 Hz. However, with the rapid increase in PV power systems, the requirement for disconnection has become a threat to system stability. This requirement, which was added to the early grid codes, was later corrected by

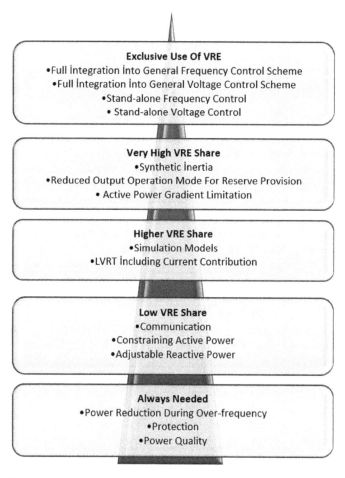

Fig. 9.16 Technical requirements according to variable renewable energy penetration

incentive programs as it was not suitable for a high share of variable RES. Disconnection of the PVs with tens of GW power in the event of high frequency can cause the system to collapse. Updates in 2012 brought a gradual power reduction in proportion to frequency, and incentives were needed to adapt existing power plants to the codes [53].

9.3.1   Fault Ride Through (FRT) Requirements

One of the most important conditions added to the grid codes with the increasing share of variable RESs is FRT. In case of faults such as short circuit, the voltage drops below its normal value until the fault is cleared. Conventional generators help to minimize the spread of voltage collapse by injecting high current into the grid and staying in the circuit in such faults.

In case of a low share of RE plants, the shutdown of plants does not affect system stability when the voltage sags below a certain value. However, when their shares in generation increase, a power loss exceeding the capacity of primary reserves may be experienced due to shutdown of many power plants. This case causes consecutive faults in the system. Therefore, most of the current grid codes require variable RE power plants to remain active for a period in the event of voltage collapse and contribute to the regulation of voltage after fault [53].

FRT requirements of PV power plants (PVPP) during fault are shown in Fig. 9.17 [56]. In the figure, there are three main regions about the voltage values during the

Fig. 9.17 FRT requirements

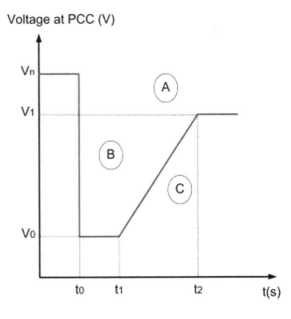

fault and duration of FRT. V_n represents nominal voltage value, V_0 indicates lower limit at voltage collapse, V_1 indicates safe voltage reached after clearing the fault. Area A is the state where PVPP works continuously. In area B, PVPP must remain connected to the system for a certain period and maintain its normal generation. It must provide maximum voltage support by adding sufficient controlled reactive current to help stabilization of voltage. In area C, PVPP does not have to stay connected to the grid, and disconnection is allowed [56].

FRT requirements have differences according to the country (Denmark, China, Spain, Romania, South Africa, USA, Germany, Turkey);

- V_0 voltage: 0–0.20 pu, t_1 time: 0.15–0.625 s.
- V_1 voltage: 0.85–0.90 pu, t_2 time: 0.5–3 s.

9.3.2   Voltage and Frequency Limits

Power system frequency is an indicator of the balance between generation and consumption. Any variation in planned generation or consumption causes system frequency to deviate from its nominal value. A sudden increase in load sets off a decrease in frequency, and generation must be increased by primary control to reach its nominal value again. Low-frequency values are the result of unexpected shutdowns of generation plants. On the other hand, excessive frequency is caused by a decrease in load or an unexpected increase in generation.

The grid codes require that the PVPPs are operated continuously within voltage and frequency values that constitute the normal operation conditions. Besides, it is requested that plants remain in operation for a certain period outside these conditions [57]. Voltages and frequency deviations outside the given tolerances and VRE generators are expected to ride through, are shown in Fig. 9.18. They are assigned as high-voltage ride through (HVRT), low-voltage ride through (LVRT), and frequency ride through (FRT) requirements [58].

In Denmark, voltage fluctuations (0.9–1.1 pu) of up to ±10% at grid connection point are normal operation conditions. The normal operation frequency is in range of 47–52 Hz for PVPPs and 49.5–50.2 Hz for WPPs. The minimum duration of FRT for PVPPs is not specified [59, 60]. The normal operation voltage for Spain is in the range of 90–111.5% of specified levels. The nominal operation frequency is in range 47.5–51.5 Hz as in Germany. According to voltage and frequency values, the minimum FRT duration is 3 s below 48 Hz [61]. The frequency range of normal operation is between 49.5 and 50.2 Hz for China. The PVPP must also withstand frequencies of 48–49.5 Hz for 10 min. If the frequency is higher than 50.2 Hz, PVPP must remain connected for 2 min, after which it must shut down [56].

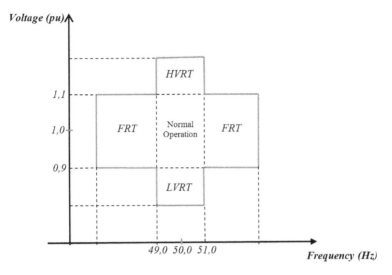

Fig. 9.18 Typical operation ranges (for a 50 Hz system)

9.3.3 *Active Power and Frequency Control*

Active power control is the ability of RESs to adjust their active power to a specified level by a specific ramping ratio. These requirements are required to ensure that system frequency remains constant to prevent overloading of transmission lines and to reduce the effects of the dynamic operation of the RESs [57].

Active power control in WPPs and PVPPs must be able to keep up with variability in energy sources. Active power control is presented as three main requirements. These are absolute power restriction, delta power restriction, and ramp rate restriction (Fig. 9.19) [56].

The absolute power limitation is used to keep power at connection point at maximum power limit determined by set values transmitted by the operator. The main reason for this is to prevent network overloading in critical situations. Absolute generation in the USA is known as power outage [56, 59, 60]. Delta power constraint is used to limit the output power of the plant to the desired value below its current active power. In this way, it is possible to reserve upwards if necessary and is a percentage of the absolute power that PVPP can normally provide. This reserve helps to have future control of PVPP when frequency or voltage deviation occurs [56, 60]. Ramp rate constraint determines the change rate of active power at power output changes or required reference values. Thus, rapid active power changes that can affect system stability are prevented. PVPP must meet these requirements despite radiation changes and cloud coverage [56].

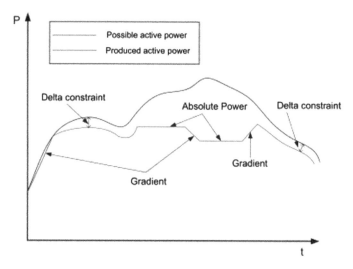

Fig. 9.19 Active power control constraints for PVPPs

9.3.4 Reactive Power and Voltage Control

With adaptations made in grid codes, the RESs need to contribute to voltage regulation of system such as conventional power plants. Power plants must be capable
of exchanging reactive power with the system to adjust the voltage at the point of
common coupling. Under normal operation conditions, the voltage level at the point
of common coupling can be increased by injecting reactive power into the system or
decreased by drawing reactive power from the system [57].

There are two main difficulties in control of the connection voltage of RESs to
the grid. First, the voltage must be kept in the dead band specified by the system
operator. The second is the obligation of the plant to comply with the operator-
specified capability curve between reactive power and active power. There are several
methods for voltage control. These are voltage control, power factor control, and
reactive power control. The voltage regulation controls voltage value depended on
the "droop" function. Here, "droop" is voltage variation that occurred due to reactive
power change. Power factor regulation controls reactive power depending on active
power value. Finally, reactive power control regulates reactive power at connection
point independently of active power. Power plants should be able to control reactive
power and voltage by the methods mentioned above [56].

9.4 Providing Flexibility

The flexibility requirements on the supply side of the power system can be met by
part-loaded operation of power plants connected to the system, load following, and

quick start/shutdown time. In cases of the load increment, part-loaded plants can contribute as load following and reserve. However, their contribution depends on ramp-up rates and the capacity difference between their maximum capacity and their actual generation. Likewise, in cases of load reduction, their contribution as load following and reserve also depends on ramp-down rates and the difference between their actual generation and minimum stable generation [12]. The evaluation of flexible and inflexible plants determined according to the variables mentioned above is given with dimensions presented in the IEA report [62], and typical characteristics of existing generation plants are shown in Table 9.3 [62]. Technologies that are providing flexibility to the power system should be evaluated with these dimensions. This case leads to different profiles for technologies, and it is necessary to consider whether a particular technology is useful for integrating the high share of VRE and how it is being evaluated. The fuel type may not be an indicator that a plant is flexible. The design characteristics of different gas and coal plants can lead to very different performance profiles. A flexible coal plant can provide lower minimum generation levels and better ramp capability than an inflexible combined-cycle gas turbine (CCGT).

Similarly, different turbine types in hydroelectric power plants can perform differently in terms of providing flexibility. However, safety regulations in nuclear power plants may block generation. At the same time, start-up times of nuclear plants are two hours from hot state to two days.

Table 9.3 Characteristics of flexible and inflexible plants according to IEA

Technology	Mini stable output (%)	Ramp rate (%/min)	Lead time, warm (h)
Gas CCGT inflexible	40–50	0.8–6	2–4
Gas CCGT flexible	15–30	6–15	1–2
Steam turbine (gas/oil)	10–50	0.6–7	1–4
Coal inflexible	40–60	0.6–4	5–7
Coal flexible	20–40	4–8	2–5
Lignite	40–60	0.6–6	2–8
Nuclear inflexible	100	0	–
Nuclear flexible	40–60	0.3–5	–

Fig. 9.20 Demand-side
management classes

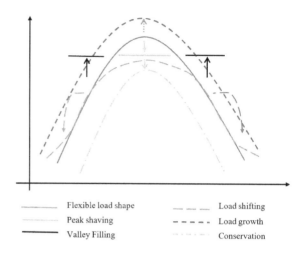

—————— Flexible load shape   — — — — Load shifting

·············· Peak shaving   – – – – – Load growth

—————— Valley Filling   ············· Conservation

9.4.1 Demand-Side Management

In a power system, many factors in the supply-side will increase flexibility, and the demand side can also contribute to flexibility [63]. The more special case of demand-side management is the demand participation, i.e., is the ability of the end consumer to control their devices and reprogram their operation [64]. It can be classified as increasing (valley filling, load growth), reducing (peak shaving, conservation), or re-scheduling (load shifting) of electricity demand (Fig. 9.20) [65].

The system operators ensure that the period in which demand is increasing coincides with the time of high generation from RESs. With this method, electricity consumption is not reduced; only the consumption is shifted to a more convenient time in terms of network operation [64]. In this way, demand-side management acts as a reserve. This condition is especially evident when low demand time and high generation time are the same [66].

When peak load and VRE generation are high, a decrease in VRE output power is a major problem for the system. In this case, the system can be eased by reducing consumption with demand-side management [67]. However, Strbac's research [66] stated that demand-side management could compete with conventional methods about supplying reserves in a system contains a high amount of unpredictable wind energy with only inflexible generation. The contribution of demand-side management is lower in a system that contains flexible plants.

9.4.2 Flexible Coal, Natural Gas, and Nuclear Plants

Coal plants are used as base load power plants, which will operate for maximum time with constant output power throughout the world. While the flexibility level of the existing coal plants can be increased by renewing used technology, new plants can be designed more flexible [68]. The undesirable effects of gained flexibility on coal power plants are low efficiency resulting from continuous start–stop and ramping, increased cost, shorter equipment life, and more maintenance requirement. Manufacturers are also looking for new solutions to mitigate these effects [28].

Nuclear power plants are base load plants with the least flexibility, according to many groups. The majority are designed to operate at full power and to be stopped only for fuel replacement or periodic maintenance. However, flexibility can be gained to these power plants with the necessary design and operation style [8, 69]. According to the International Atomic Energy Agency, most of the existing nuclear power plants have a power output range of 50–100% of reactor thermal power and ramping rates of up to 5%/min. However, these features are not a part of their daily operations [69]. Only certain countries have experience in designing and operating in a wide flexibility range of nuclear power plants. In France, some nuclear power plants can change their output power from 30 to 100% in 1 h and from 60 to 100% in 30 min under load tracking mode [8, 69]. The USA has increased the installed capacity of pumped hydraulic power plants to solve this flexibility problem. When the demand is low, a part of the generation of nuclear power plants is directed to these plants so that stored energy is used when demand is high [70].

With natural gas power plants, which are used as base load or medium-level power plants, flexibility can be gained to the system at a certain rate. The most common are CCGTs due to their capacity diversity, high efficiency, and low energy costs. The new generation of high-performance CCGTs is much faster than conventional CCGTs with a start-up time of 40 min. However, gained flexibility causes wear on mechanical components, as in other plants, it requires more frequent maintenance and increases operating costs [71].

9.4.3 Flexible Cogeneration Plants

Cogeneration plants are important to ensure high RES penetration rates. Correct heating and cooling applications can be flexibility sources. However, cogeneration plants are not flexible due to their generation planned for heat load in many countries [62]. Flexibility can be given to cogeneration plants by changing operations and equipment [72, 73]. Changes in the operating methods according to heat load, electricity demand, or existing generation of cogeneration plants or use of heat storage tanks with electric boilers can be given as examples [73].

Denmark is one of the countries that have the largest cogeneration system in Europe, with a 50% rate in electricity generation [74]. Furthermore, cogeneration plants are operated flexibly in the country with high WE penetration. These operating modes are shown in Fig. 9.21 [62].

Cogeneration plants use fossil fuels to meet demand in case of high heat demand and medium/low generation from RES (Mod 1). In the case of the high generation from RES and low heat demand, the output of the cogeneration plant can be reduced.

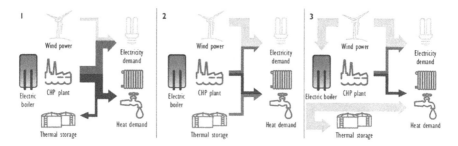

Fig. 9.21 Operating modes of wind power and cogeneration plants in Denmark

A part of the demand can be met by heat storage if necessary (Mod 2). In cases of generation from RESs exceeds demand, electrical energy can be used in electric boilers to meet heat demand, can be stored as heat, or can be used for both. (Mod 3) [62].

9.4.4 Renewable Energy Generation Curtailment

Due to the variability and predictability issues of variable generation, the integration of wind and solar power may be limited. When wind power and PV systems cause transmission or operational constraints, the system operator may be forced to accept less wind and solar power than what is available. This event is called curtailment [75]. The low capacity factor of WPP affects transmission line design. Due to their variable structure, the time of usage at installed power is limited in their integration [76]. Therefore, the renewal of the transmission line for the entire installed capacity is often not an economical method. Therefore, if there are supply constraints, generation curtailment can be a more economical solution [75, 76].

In addition, congestion due to insufficient transmission line capacity or regional network restrictions requires system operators to use resources at higher prices. Due to differences in construction times, VREs can be built before required transmission line projects. So, their generation may be curtailment until the transmission line infrastructure is commissioned. In this case, generation curtailment at regular intervals can be a more economical solution than expanding the transmission line.

The minimum output power of thermal power plants is also one of the reasons for generation curtailment. If VREs cannot meet the downward reserve requirement of the system, thermal power plants undertake this task. To provide reserves, power plants increase their power levels and reduce the share of generation from RE. However, this is not a problem for modern wind and solar power plants. In addition to thermal power plants, hydroelectric power plants may also be subjected to a limit on their output power due to environmental reasons and irrigation problems.

On the other hand, generation curtailment is also applied in the distribution system to prevent high penetration levels and feedback (generation more energy than feeder consumes). Feedback can cause voltage control problems due to the variability of RE. If the infrastructure is not designed or adapted for such an operation, it is a problem for feedback protection devices.

Finally, asynchronous generation levels need to be limited, especially in small or island systems, to maintain system frequency and stability. Modern WPP and PVs are connected to the grid through power electronic elements. Since these plants replace synchronous generators that contribute to system inertia, serious fluctuations may occur at system frequency in case of failure [75].

Many power systems around the world with high RE rates encounter generation curtailment problems. At the end of 2017, the installed capacity of RE is 619 GW in China so that is the leader. Of this value, 341 GW is hydroelectric, 164 GW is WE, and 131 GW is PV. China, the leader in renewable energy, is the country that faces the most serious problem of renewable curtailment [77]. The structural reasons for this problem are weak network, clustering of WE resources in remote areas, high coal plant ratio, and lack of adequate market mechanisms. Unfavorable tariff guarantee, unreasonable dispatch priorities, lack of required grid codes, and low wind forecast accuracy are operational challenges [78].

In China, the generation curtailment problem began in the Inner Mongolia region in 2009 and spread throughout the country in 2010. Generation curtailment in solar energy occurred in 2013. Between 2013 and 2016, there was an average of 15% solar energy generation curtailment across the country. Between 2011 and 2015, an average of 15% WE generation curtailment was across the country. This rate increased to 43% in the northern regions of China in 2016 [79]. In China, WE generation curtailment, on average, was 17% in 2016, according to data from the National Energy Administration (49.7 TWh) [80].

Wind and solar power are showing rapid growth in China. In 2018, China added a 44.26 GW solar PV. By increasing about 20 GW of WE, total installed power reached 184 GW at the end of 2018. It means that China accounts for 52.9% of wind and solar capacity additions, which now show that wind and solar are the main sources in electricity generation. Importantly, the generation from wind and solar PV continues to increase, while curtailment levels are decreasing. In 2018, wind generation increased by 20% annual, while curtailments decreased by 5 points to 7%. In the same period, the generation from solar energy increased by 50%, while curtailments decreased by 2.8 points and were 3% [81].

In 2016, an island system, Ireland and Northern Ireland, 227 GWh generation curtailment was occurred from 7620 GWh WE generation. This value was decreased by 215 GWh compared to the previous year. 52% of these curtailments were caused by problems in the system and 48% due to regional network problems. Ireland had increased the asynchronous penetration limit in the system from 50 to 55% in March 2016, and 60% started testing in November 2016 [82]. Significant improvements have been achieved in generation curtailments in Italy between 2009 and 2014. In these years, generation curtailment has not occurred in solar energy; the 10.7% curtailment in WE in 2009 decreased to 0.8% in 2014. This reduction was achieved through investments to compensate for the insufficiency of the transmission line between south and north, which is one of the main reasons for curtailment.

On the other hand, there was no significant curtailment in Denmark and Portugal that have high WE penetration. In Portugal, according to codes, curtailment from renewable generation is not allowed except for technical problems. For the 317 h in which demand was exceeded in 2016, Denmark did not need to generation curtailment by using interconnections with neighboring countries [75]. In Germany, the amount of unused energy due to generation curtailment increased by three times in 2014 and 2015 compared to the previous year. In 2015, compensation for curtailment was estimated at 478 million euros [83].

RE generation curtailment can be considered as one of the indicators of flexibility. The high amount of curtailment indicates that the system has not the flexibility to use variable RESs in its proportions. With increasing penetration levels, curtailment problems may increase, and this can be a deterrent reason for new investments.

9.4.5 Strengthening and Expanding Transmission Lines

Strengthening of the transmission line is an important factor in RESs integration to network with the contribution of increasing flexibility. These contributions are such as balancing energy generation spread over a wide area, providing access to remote regions where RE generation is intensive, facilitating exchange with neighboring countries, and connecting to international energy markets is an important factor in the integration of RESs in the network [68]. Reserves and load diversity in the planning and operation of transmission lines help to balance variable renewable generation. As the balancing region expands, the variability rate of RE decreases and thus can increase power system flexibility [84].

There are also some difficulties in RESs integration into the network [85]. Due to the nature of wind and solar, problems are caused by the fact that RES are spread over a wide geographical area. A significant generation or demand increment in a region causes uncertainties in transmission line planning. Besides, a WPP project in the remote region cannot be financed without access to the transmission line. The plan, license, and construction period of a transmission line can last up to 5–10 years. Also, the transmission line cannot be constructed without proving the necessity of the line for the operation of WPP. It is a disadvantage that the manufacturer pays for a transmission line in advance, and the inclusion of new manufacturers in this cost is a controversial topic [85, 86].

From an economic point of view, it is costlier to connect remote RESs to the network than conventional systems [85]. To operate the transmission line economically, the price difference between the high-priced zone and the low-priced zone must be higher than annual investment and operation costs. To achieve this, it is desirable to carry a large amount of energy by low-cost transmission [86].

Technically, there are problems with network topologies and connection types. Efficient and inexpensive network connection "spaghetti" occurs where each power plant in a remote region has connected the network on its own (Fig. 9.22a). The scale efficient network extension (SENE) structure (Fig. 9.22b), where a region is

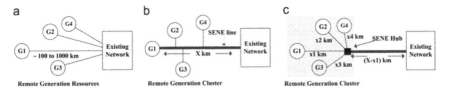

Fig. 9.22 Transmission line topologies for connection of generation in a remote area

connected to the consumption center by a high-voltage line, is more useful when power plants that can be built later are considered. Another option is created by adding a center to the SENE approach (Fig. 9.22c [85]). An additional high direct voltage (HVDC) line may be required if there are thousands of MW generation in a particular region [85].

In Germany, the intensity of generation from WE is in the north, whereas consumption intensity is in the south. This high amount of long-distance transmission causes bottlenecks in the network. Power transmission is carried out through transmission lines of neighboring countries (Poland, Czech Republic, Netherlands, and Belgium) instead of the national transmission line to eliminate this problem [87]. Malek et al. [88] pointing to this problem stated that, with the increase of wind and solar energy generation in Germany, congestion would be observed, especially in Austria and Poland networks. With the decommissioning of nuclear power plants (8386 MW by 2022) and conventional power plants, the southern generation will be significantly reduced. Additional transmission line capacity will be required in areas where conventional power plants, RES, and storages (e.g., Scandinavian countries) are located to ensure supply security in this region. Therefore, Germany plans to commission two long-distance HVDC transmission lines (Fig. 9.23) in 2025 [87–89].

The European Network of Transmission System Operators (ENTSO-E) launched a European-wide project in 2010 called the Ten Year Network Development Plan (TYNDP) with national and regional investment plans. Especially in line with targets determined by the European Union, it is aimed to provide more economic RE integration [89].

According to ENTSO-E Network Development Plan 2018 report, in 2030 scenarios, it is that 48–58% of the demands will be met from renewable energy sources. It will reduce CO_2 emissions by 65–75% compared to the 1990s. Thus, to improve the flexibility of the system, 166 transmission projects consisting of an overhead line, subsea, and underground cables have been proposed. For this, an investment of approximately EUR 114 billion is proposed. Also, the importance of interconnections between the Member States for further integration of the VRE is emphasized. It was stated that if the nominal transmission capacity of the interconnected members was below 30% of the peak load, the options for additional interconnections are required. Regarding the security of supply and RES integration criteria, the existing interconnection grid shows additional needs for interconnection development to be the most urgent in Spain, Great Britain, Ireland, Italy, Greece, Finland, France, Romania, and Poland in all scenarios [90].

Fig. 9.23 Routes for a transmission line that required strengthening of Germany [89]

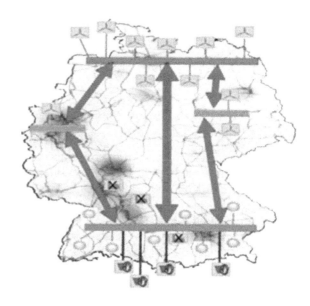

Social opposition also creates delays in the construction of new transmission lines [91]. Dynamic evaluation of transmission capacity and better control of power flow and more efficient use of existing lines are an alternative solution. Nearly, all of the transmission line capacity is used with the equipment providing power flow control such as phase changer, HVDC transmission lines, and flexible alternating current transmission systems (FACTS). Consequently, strengthening of the transmission line is inevitable at higher penetration levels [92].

9.5 Conclusions

The main purpose of power system operation is to balance supplied energy with electricity demand. Short, medium and long-term targets in order to achieve this purpose are as listed below:

Ensuring power quality, i.e., voltage and frequency stability, (from milliseconds to minutes),

The generation must meet planned demand for a certain time, and generated electrical energy must be provided to reach load (from minutes to hours),

The ability of generation and transmission capacity will be to meet demand in all regions of the system, forming necessary generation centers and transmission lines according to demand (from weeks to seasons).

This part of the book explains the flexibility requirement of the power system to accomplish the main purposes of the system regarding the aforementioned targets, while the share of variable renewable energy sources, especially PVPPs, is increasing in the generation side of the power system.

With the increase in VRE investments, power system flexibility has taken a different dimension, and a power system transformation has taken place. With the increase in variability and uncertainty rates, conventional flexibility has left its place to a concept that encompasses all resources enabling efficient and reliable power system operation. With costs decrease and the increase in VRE investments in the world, the expansion of solar energy systems is accelerating.

In this study, tests for extra-high voltage lines were performed on an internationally validated simulation program. The results obtained for different PVPP penetration rates show that voltage values at all penetration levels are within normal operation condition limits, i.e., there is no voltage unbalance. On the other hand, when the frequency values are examined, the normal operation band range of large-scale power systems is exceeded in lack of VRE-based generation at 30% penetration level. This is realized for the normal operation band of small-scale power systems at 20% penetration.

Grid codes are required for the successful integration of VRE. For power systems operation, VRE generation causes challenges that contrast with conventional generation. As response to these challenges, VRE is rapidly evolving with the provision of technologies such as generation, storage and control systems, and ease of new operational applications for power systems. These technological and operational advances provide greater flexibility for electrical network operation and greater VRE share. While new technologies are being used and new operational practices are adopted, some rules need to be set for all actors in order to guarantee ensuring electrical service for consumers. For successful integration of VRE into the grid, all three components need to be considered: technology, operation, and regulation. Existing grid codes have traditionally been developed to allow power plants connection to grid according to synchronous generators. In order to avoid any problems related with electrical system operation, it has become more important to develop grid codes for power plants that use renewable generation as their main source. Studies about grid codes are carried out on four headings as (a) fault ride through requirements, (b) voltage and frequency deviation boundaries, (c) active power and frequency control, and (d) voltage and reactive power control.

Power systems flexibility is not limited only power plant operation, so the solution is not just generation flexibility. In order to ensure voltage and frequency stability of power system, demand-side management, flexible operation of power plants, generation curtailment, strengthening and expanding transmission lines solutions are required in addition to loading and load shedding process in the generation stage.

In particular, two headings stand out from the solution proposals mentioned above. The first of these is interconnects that make VRE integration into the energy system easy and cost-effective as reliably. Interconnections allow the resources of flexibility to be shared among different countries. A wider geographic area will also soften VRE generation variability and reduce uncertainty. The second is the use of storage systems that ensure maximum existing energy usage. Storage systems are also very open to development in terms of both the technological and usage of generation centers for storage.

References

1. International Energy Agency: Empowering Variable Renewables Options for Flexible Electricity Systems. Paris (2008)
2. Nosair, H., Bouffard, F.: Flexibility envelopes for power system operational planning. IEEE Trans. Sustain. Energy **6**, 800–809 (2015)
3. Denholm, P., Hand, M.: Grid flexibility and storage required to achieve very high penetration of variable renewable electricity. Energy Policy **39**, 1817–1830 (2011)
4. Lannoye, E., Flynn, D., O'Malley, M.: Evaluation of power system flexibility. IEEE Trans. Power Syst. **27**, 922–931 (2012)
5. Lannoye, E., Flynn, D., O'Malley, M.: The role of power system flexibility in generation planning. In: 2011 IEEE Power and Energy Society General Meeting, pp 1–6 (2011)
6. Dvorkin, Y., Kirschen, D.S., Ortega-Vazquez, M.A.: Assessing flexibility requirements in power systems. IET Gener. Transm. Distrib. **8**, 1820–1830 (2014)
7. Energiewirtschaftliches Institut an Der Universität Zu Köln: Flexibility Options in European Electricity Markets in High Res-E Scenarios. Study on Behalf of the International Energy Agency, Cologne (2012)
8. International Energy Agency: Harnessing Variable Renewables: A Guide to the Balancing Challenge. OECD Publishing, Paris (2011)
9. Kehler, J.H., Hu, M.: Planning and operational considerations for power system flexibility. In: 2011 IEEE Power and Energy Society General Meeting, pp. 1–3 (2011)
10. Bouffard, F., Ortega-Vazquez, M.: The value of operational flexibility in power systems with significant wind power generation. In: 2011 IEEE Power and Energy Society General Meeting, pp. 1–5 (2011)
11. Bucher, M.A., Delikaraoglou, S., Heussen, K., Pinson, P., Andersson, G.: On quantification of flexibility in power systems. In: 2015 IEEE Eindhoven PowerTech, pp. 1–6 (2015)
12. Ma, J., Silva, V., Belhomme, R., Kirschen, D.S., Ochoa, L.F.: Evaluating and planning flexibility in sustainable power systems. In: 2013 IEEE Power and Energy Society General Meeting, pp 1–11 (2013)
13. Holttinen, H., Tuohy, A., Milligan, M., Lannoye, E., Silva, V., Müller, S., Sö, L.: The flexibility workout: managing variable resources and assessing the need for power system modification. IEEE Power Energy Mag. **11**(6), 53–62 (2013)
14. Kundur, P., Paserba, J., Ajjarapu, V., Andersson, G., Bose, A., Canizares, C., Hatziargyriou, N., Hill, D., Stankovic, A., Taylor, C.: Definition and classification of power system stability IEEE/CIGRE joint task force on stability terms and definitions. IEEE Trans. Power Syst. **19**, 1387–1401 (2004)
15. Tielens, P., Van Hertem, D.: The relevance of inertia in power systems. Renew. Sustain. Energy Rev. **55**, 999–1009 (2016)
16. Seneviratne, C., Ozansoy, C.: Frequency response due to a large generator loss with the increasing penetration of wind/Pv generation–a literature review. Renew. Sustain. Energy Rev. **57**, 659–668 (2016)
17. Hsieh, E., Anderson, R.: Grid flexibility: the quiet revolution. The Electr. J. **30**, 1–8 (2017)
18. Papaefthymiou, G., Grave, K., Dragoon, K.: Flexibility Options in Electricity Systems. ECOFYS, Berlin (2014)
19. Makarov, Y.V., Loutan, C., Ma, J., De Mello, P.: Operational impacts of wind generation on California power systems. IEEE Trans. Power Syst. **24**, 1039–1050 (2009)
20. Ela, E., Milligan, M., Bloom, A., Botterud, A., Townsend, A., Levin, T., Frew, B.A.: Wholesale electricity market design with increasing levels of renewable generation: incentivizing flexibility in system operations. The Electr. J. **29**, 51–60 (2016)

21. National Renewable Energy Laboratory: Flexibility in 21st Century Power Systems. https://www.nrel.gov/docs/fy14osti/61721.pdf (2014). Accessed 01 Dec 2019
22. International Energy Agency: Status of Power System Transformation 2018 Advanced Power Plant Flexibility. Paris (2018)
23. North American Electric Reliability Corporation: Flexibility Requirements and Potential Metrics for Variable Generation: Implications for System Planning Studies. Princeton, NJ (2010)
24. Lee, C.T., Hsu, C.W., Cheng, P.T.: A low-voltage ride-through technique for grid-connected converters of distributed energy resources. IEEE Trans. Ind. Appl. **47**, 1821–1832 (2011)
25. Yang, L., Xu, Z., Ostergaard, J., Dong, Z.Y., Wong, K.P.: Advanced control strategy of Dfig wind turbines for power system fault ride through. IEEE Trans. Power Syst. **27**, 713–722 (2012)
26. Alizadeh, M., Moghaddam, M.P., Amjady, N., Siano, P., Sheikh-El-Eslami, M.: Flexibility in future power systems with high renewable penetration: a review. Renew. Sustain. Energy Rev. **57**, 1186–1193 (2016)
27. Troy, N., Denny, E., O'Malley, M.: Base-load cycling on a system with significant wind penetration. IEEE Trans. Power Syst. **25**(2), 1088–1097 (2010)
28. Henderson, C.: Increasing the Flexibility of Coal-Fired Power Plants. IEA Clean Coal Centre, London (2014)
29. Lefton, S.A., Besuner, P., Grimsrud, G., Strauss, S.: Understand what it really costs to cycle fossil-fired units. Power **141**, 41–46 (1997)
30. Van den Bergh, K., Delarue, E.: Cycling of conventional power plants: technical limits and actual costs. Energy Convers. Manag. **97**, 70–77 (2015)
31. International Energy Agency: Solar Energy Mapping the Road Ahead. Paris (2019)
32. Vittal, V.: The impact of renewable resources on the performance and reliability of the electricity grid. The Bridge **40**, 5–12 (2010)
33. Hecker, L., Zhou, Z., Osborn, D., Lawhorn, J.: Value based transmission planning process for joint coordinated system plan. In: 2009 IEEE Power and Energy Society General Meeting, pp 1–6 (2009)
34. Neville, A.: Top Plant: Desoto Next Generation Solar Energy Center, Desoto County, Florida. Power **154**, 32 (2010)
35. Romero-Cadaval, E., Francois, B., Malinowski, M., Zhong, Q.C.: Grid-connected photovoltaic generation plants as alternative energy sources. IEEE Ind. Electron. Mag. **9**, 18–32 (2015)
36. Shah, R., Mithulananthan, N., Bansal, R., Ramachandaramurthy, V.: A review of key power system stability challenges for large-scale Pv integration. Renew. Sustain. Energy Rev. **41**, 1423–1436 (2015)
37. Zhang, Y., Zhu, S., Sparks, R., Green, I.: Impacts of solar Pv generators on power system stability and voltage performance. In: 2012 IEEE Power and Energy Society General Meeting, pp. 1–7 (2012)
38. Eftekharnejad, S., Vittal, V., Heydt, G.T., Keel, B., Loehr, J.: Impact of increased penetration of photovoltaic generation on power systems. IEEE Trans. Power Syst. **28**, 893–901 (2013)
39. Tamimi, B., Cañizares, C., Bhattacharya, K.: System stability impact of large-scale and distributed solar photovoltaic generation: the case of Ontario, Canada. IEEE Trans. Sustain. Energy **4**(3), 680–688 (2013)
40. Liu, H., Jin, L., Le, D., Chowdhury, A.: Impact of high penetration of solar photovoltaic generation on power system small signal stability. In: 2010 International Conference Power System Technology, IEEE, Hangzhou, 24–28 Oct 2010
41. Ravichandran, S., Dasan, S.B., Devi, R.K.: Small signal stability analysis of grid connected photo voltaic distributed generator system. In: 2011 International Conference on Power and Energy Systems, IEEE, Chennai, 22–24 Dec 2011
42. Du, W., Wang, H., Xiao, L.: Power system small-signal stability as affected by grid-connected photovoltaic generation. Int. Trans. Electr. Energy Syst. **22**, 688–703 (2012)
43. Eftekharnejad, S., Vittal, V., Heydt, G.T., Keel, B., Loehr, J.: Small signal stability assessment of power systems with increased penetration of photovoltaic generation: a case study. IEEE Trans. Sustain. Energy **4**, 960–967 (2013)

44. Alquthami, T., Ravindra, H., Faruque, M., Steurer, M., Baldwin, T.: Study of photovoltaic integration impact on system stability using custom model of Pv arrays integrated with PSS/E. In: North American Power Symposium, IEEE, Arlington, 26–28 Sept 2010
45. Abdlrahem, A., Venayagamoorthy, G.K., Corzine, K.A.: Frequency stability and control of a power system with large Pv plants using Pmu information. In: North American Power Symposium (NAPS), IEEE, Manhattan 22–24 Sept 2013
46. Kroposki, B., Johnson, B., Zhang, Y., Gevorgian, V., Denholm, P., Hodge, B.M., Hannegan, B.: Achieving a 100% renewable grid: operating electric power systems with extremely high levels of variable renewable energy. IEEE Power Energ. Mag. **15**, 61–73 (2017)
47. Liu, Y., Choi, S., Meliopoulos, A.S., Fan, R., Sun, L., Tan, Z.: Dynamic state estimation enabled preditive inverter control. In: 2016 IEEE Power and Energy Society General Meeting, IEEE, Boston, 17–21 July 2016
48. DIgSILENT: Digsilent Powerfactory 2018 User Manual. DIgSILENT GmbH, Gomaringen (2018)
49. Çiftkaya B.: Investigation of heavy duty gas turbines and their simulation. M.Sc. Thesis, Istanbul Technical University (2013)
50. Kundur, P.: Power System Stability and Control. McGraw-Hill, New York (1994)
51. Zea, A.A.: Power system stabilizers for the synchronous generator-tuning and performance evaluation. M.Sc. Thesis, Chalmers University of Technology (2013)
52. Oguz, G.: Voltage stability control on power systems by fuzzy logic. M.Sc. Thesis, Istanbul Technical University (2004)
53. Ackermann, T., Martensen, N., Brown, T., Schierhorn, P., Boshell, F., Gafaro, F., Ayuso, M.: Scaling up Variable Renewable Power: The Role of Grid Codes. International Renewable Energy Agency, Abu Dhabi (2016)
54. European Network of Transmission System Operators for Electricity: ENTSO-E Network Code for Requirements for Grid Connection Applicable to all Generators. Brussels (2013)
55. International Electrotechnical Commission: Grid Integration of Large Capacity Renewable Energy Sources and Use of Large-Capacity Electrical Energy Storage. Geneva (2012)
56. Cabrera-Tobar, A., Bullich-Massagué, E., Aragüés-Peñalba, M., Gomis-Bellmunt, O.: Review of advanced grid requirements for the integration of large scale photovoltaic power plants in the transmission system. Renew. Sustain. Energy Rev. **62**, 971–987 (2016)
57. Sourkounis, C., Tourou, P.: Grid code requirements for wind power integration in Europe. In: Power Options for the Eastern Mediterranean Region, Limassol, 19–21 Nov 2012
58. Energy Sector Management Assistance Program: Grid Integration Requirements For Variable Renewable Energy Technical Guide. World Bank, Washington, DC (2019)
59. Energinet, D.K.: Technical Regulation 3.2.2 for PV Power Plants Above 11 kW. Fredericia (2016)
60. Energinet, D.K.: Technical Regulation 3.2.5 for Wind Power Plants Above 11 kW. Fredericia (2016)
61. Spanish Wind Energy Associations: Operation Procedure O.P. 12.2: Technical Requirements for Wind Power and Photovoltaic Installations and Any Generating Facilities Whose Technology Does Not Consist on a Synchronous Generator Directly Connected to the Grid. Madrid (2008)
62. International Energy Agency: The Power Transformation Wind, Sun and the Economics of Flexible Power System. Paris (2014)
63. Kirschen, D.S., Rosso, A., Ma, J., Ochoa, L.F.: Flexibility from the demand side. In: 2012 IEEE Power and Energy Society General Meeting, pp. 1–6 (2012)
64. Pina, A., Silva, C., Ferrão, P.: The impact of demand side management strategies in the penetration of renewable electricity. Energy **41**, 128–137 (2012)
65. Lund, P.D., Lindgren, J., Mikkola, J., Salpakari, J.: Review of energy system flexibility measures to enable high levels of variable renewable electricity. Renew. Sustain. Energy Rev. **45**, 785–807 (2015)
66. Strbac, G.: Demand side management: benefits and challenges. Energy Policy **36**, 4419–4426 (2008)

67. Moura, P.S., De Almeida, A.T.: The role of demand-side management in the grid integration of wind power. Appl. Energy **87**, 2581–2588 (2010)
68. Martinot, E.: Grid integration of renewable energy: flexibility, innovation, and experience. Annu. Rev. Environ. Resour. **41**, 223–251 (2016)
69. International Atomic Energy Agency: Non-Baseload Operation in Nuclear Power Plants: Load Following and Frequency Control Modes of Flexible Operation. Vienna (2018)
70. International Energy Agency: Large-Scale Electricity Interconnection-Technology and Prospects for Cross-Regional Networks. Paris (2016)
71. World Bank Group: Bringing Variable Renewable Energy Up to Scale: Options for Grid Integration Using Natural Gas and Energy Storage. Washington, DC (2015)
72. Beer, M., Huber, M., Mauch, W.: Flexible operation of cogeneration plants-chances for the integration of renewables. In: 11th IAEE European Conference Energy Economy, Policies and Supply Security: Surviving the Global Economic Crisis, Vilnius, 25–28 Aug 2010
73. Chen, X., Kang, C., O'Malley, M., Xia, Q., Bai, J., Liu, C., Sun, R., Wang, W., Li, H.: Increasing the flexibility of combined heat and power for wind power integration in China: modeling and implications. IEEE Trans. Power Syst. **30**, 1848–1857 (2015)
74. Cogeneration Observatory and Dissemination Europe: D5.1—Final Cogeneration Roadmap Member State: Denmark. http://www.code2-project.eu/wp-content/uploads/Code-2-D5-1-Final-non-pilor-Roadmap-Denmark_f2.pdf (2014). Accessed 20 Sept 2019
75. Bird, L., Lew, D., Milligan, M., Carlini, E.M., Estanqueiro, A., Flynn, D., Gomez-Lazaro, E., Holttinen, H., Menemenlis, N., Orths, A.: Wind and solar energy curtailment: a review of international experience. Renew. Sustain. Energy Rev. **65**, 577–586 (2016)
76. Burke, D.J., O'Malley, M.J.: Factors influencing wind energy curtailment. IEEE Trans. Sustain. Energy **2**, 185–193 (2011)
77. International Renewable Energy Agency (IRENA): Renewable Capacity Statistics 2018, Abu Dhabi (2018)
78. Li, C., Shi, H., Cao, Y., Wang, J., Kuang, Y., Tan, Y., Wei, J.: Comprehensive review of renewable energy curtailment and avoidance: a specific example in China. Renew. Sustain. Energy Rev. **41**, 1067–1079 (2015)
79. Zhang, S., Andrews-Speed, P., Li, S.: To what extent will China's ongoing electricity market reforms assist the integration of renewable energy. Energy Policy **114**, 165–172 (2018)
80. Global Wind Energy Council: Global Wind Report. Brussels (2016)
81. International Energy Agency: China Power System Transformation Assessing the Benefit of Optimised Operations and Advanced Flexibility Options. Paris (2019)
82. EirGrid and SONI: Annual Renewable Energy Constraint and Curtailment Report 2016. Dublin (2017)
83. Schermeyer, H., Vergara, C., Fichtner, W.: Renewable energy curtailment: a case study on today's and tomorrow's congestion management. Energy Policy **112**, 427–436 (2018)
84. Miller, M., et al.: Status Report on Power System Transformation: A 21st Century Power Partnership Report. National Renewable Energy Laboratory, Golden (2015)
85. Hasan, K.N., Saha, T.K., Eghbal, M., Chattopadhyay, D.: Review of transmission schemes and case studies for renewable power integration into the remote grid. Renew. Sustain. Energy Rev. **18**, 568–582 (2013)
86. Smith, J.C., et al.: Transmission planning for wind energy in the United States and Europe: status and prospects. Wiley Interdiscip. Rev.: Energy Environ. **2**, 1–13 (2013)
87. Brown, T.: Transmission network loading in Europe with high shares of renewables. IET Renew. Power Gener. **9**, 57–65 (2014)
88. Málek, J., Rečka, L., Janda, K.: Impact of German Energiewende on transmission lines in the Central European region. Energ. Effi. **11**, 683–700 (2018)
89. European Network of Transmission System Operators for Electricity: Ten-Year Network Development Plan 2016 Project Sheets. https://docstore.entsoe.eu/Documents/TYNDP%20documents/TYNDP%202016/projects/TYNDP2016-project-sheets.pdf (2014). Accessed 15 Oct 2019

90. European Network of Transmission System Operators for Electricity: Connecting Europe: Electricity 2025–2030–2040. Brussels (2019)
91. Cain, N.L., Nelson, H.T.: What drives opposition to high-voltage transmission lines. Land Use Policy **33**, 204–213 (2013)
92. Papaefthymiou, G., Dragoon, K.: Towards 100% renewable energy systems: uncapping power system flexibility. Energy Policy **92**, 69–82 (2016)

Chapter 10
Photovoltaic Plant Output Power Forecast by Means of Hybrid Artificial Neural Networks

E. Ogliari and A. Nespoli

Abstract The main goal of this chapter is to show the set up a well-defined method to identify and properly train the hybrid artificial neural network both in terms of number of neurons, hidden layers and training set size in order to perform the day-ahead power production forecast applicable to any photovoltaic (PV) plant, accurately. Therefore, this chapter has been addressed to describe the adopted hybrid method (PHANN— Physic Hybrid Artificial Neural Network) combining both the deterministic clear sky solar radiation algorithm (CSRM) and the stochastic artificial neural network (ANN) method in order to enhance the day-ahead power forecast. In the previous works, this hybrid method had been tested on different PV plants by assessing the role of different training sets varying in the amount of data and number of trials, which should be included in the "ensemble forecast." In this chapter, the main results obtained by applying the above-mentioned procedure specifically referred to the available data of the PV power production of a single PV module are presented.

Keywords Artificial Neural Networks · Day-ahead forecast · Computational Intelligence · PHANN

10.1 Introduction

In the last decades, power production from renewable energy sources has spread around the world, due to the increasing demand for electricity and the necessity to progressively phase out from traditional technologies, based on the exploitation of fossil fuels. Global warming and the progressive reduction of fossil fuels availability have brought nations to sign global contracts in order to diminish the emissions of greenhouse gases at worldwide level [1].

In particular, photovoltaic plants, among the renewable energy sources (RESs), constitute the reference technologies on which governments leverage to achieve this

E. Ogliari (✉) · A. Nespoli
Department of Energy, Politecnico di Milano, 20133 Milan, Italy
e-mail: emanuelegiovanni.ogliari@polimi.it

© The Editor(s) (if applicable) and The Author(s), under exclusive license to Springer Nature Switzerland AG 2020
A. Mellit and M. Benghanem (eds.), *A Practical Guide for Advanced Methods in Solar Photovoltaic Systems*, Advanced Structured Materials 128,
https://doi.org/10.1007/978-3-030-43473-1_10

shift to renewable energy. As a drawback, these sources are affected by a high level of uncertainty and variability, introducing, therefore, new difficulties in ensuring grid stability and managing their production. For this reason, nowadays, the demand for reliable forecasting methods has increased significantly, leading the researchers to experiment new methods and procedures, in particular related to the forecast of solar plants production, that provided in 2017 2% of the global power output, with 398 GW of installed capacity generating over 460 TWh of energy [2]. For the future years, renewable power capacity is set to expand by 50% between 2019 and 2024, led by solar PV. This increase of 1200 GW is equivalent to the total installed power capacity of the USA today [3]. Solar PV alone accounts for almost 60% of the expected growth, with onshore wind representing one-quarter [4].

The main goal of the work performed was to identify a procedure to improve the PV power forecast accuracy by means of an hybrid artificial neural network. The analysis has been based on the preprocessed data collected at the Solar Tech Lab, at Politecnico di Milano, during the years 2017 and 2018. An hybrid ANN has been set up on these available data in order to obtain a set of 40 forecasts (trials) with the aim to perform the post-processing work of ensemble forecast.

10.2 Photovoltaic Power Forecast

The prediction of energy, which was often applied to the electric loads, is a typical application of time series analysis methods. Instead, the energy supply forecasting is a more recent topic which is closely related to the spreading of RES and their unpredictable weather dependency. For instance, one of the major sources of variability in photovoltaic power production is related to moving clouds and solar position, together with weather systems [5]. The variability associated to solar position is completely deterministic and can be computed by adopting suitable equations and programs, while the uncertainty associated to clouds is mostly stochastic, because the existent models are not able to predict precisely their time evolution since it is affected by too many variables, and their dynamics have not been modeled exhaustively yet. In addition, the accuracy of the forecast strongly depends on the variability of the weather conditions which are occurring in a given location and this may change from year to year.

Consequently, the models for power forecasting can be classified according to different criteria: forecast horizon, model type and forecasting technique [6].

In recent years, several power forecasting models related to PV plants have been published. The existing solutions can be classified into the categories of physical, statistical and hybrid method [7] as depicted in Fig. 10.1 [8]. Some of these models were at first oriented to obtain solar radiation predictions [9, 10], while other works present models specifically dedicated to the forecasting of the hourly power output from PV plants [11, 12]. Nowadays, the most applied techniques to model the stochastic nature of solar irradiance at the ground level and thus the power output of PV installations are the statistical methods. In particular, regression methods are often employed to

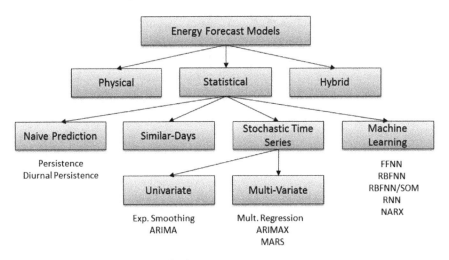

Fig. 10.1 Energy forecasting classification

describe complex nonlinear atmospheric phenomena for few-hours ahead forecast and specific soft computing techniques based on artificial neural network (ANN) are used for few-hours power output forecast [13]. Some other papers use physical methods reporting the comparison of the results obtained with different models based on two or more forecasting techniques [14]. Nowadays, the most important forecasting horizon is 24 h of the next days. Only a few papers describe forecasting models which used to predict the daily irradiance or directly energy production of the PV plant for all the daylight hours of the following day [14–16].

RES energy production forecasting methods are commonly divided in different categories: physic, stochastic and hybrid. An analysis of the state-of-the-art approaches is proposed in [8]. Physical models are based on mathematical equations which describe the ability of PV systems to convert the introduced meteorological resources into electrical power [9, 14]. These models can be very simple, if based only on the global solar radiation, or more complicated if they include additional parameters. In fact, it is not easy to predict PV module energy production since it depends on several parameters. The conversion process is affected by solar radiation, cell temperature, the presence of shadow [17] and the load resistance. Moreover, information provided by manufacturer is usually limited and only at nominal operating conditions. The major disadvantage of these models is that they have to be designed specifically for a particular plant and location.

Statistical methods are based on the concept of persistence or stochastic time series. When developing these approaches, it is assumed that an algebraic link exists among the parameters affecting the plant power production. These links can be used to forecast future values or patterns based on the previous values of the time series under consideration and/or other time series variables. The stochastic category of solar forecasting methods includes data-driven techniques that are developed by fitting the parameters of the model function in a training phase with input and target data. Regression methods often employed to describe complex nonlinear atmospheric phenomena include the autoregressive moving averages (ARMAs) method, as well as its variations, such as the autoregressive integrated moving averages (ARIMAs) method [10, 13]. The performance of these models is very good for few-minutes to few-hours ahead forecasts [10, 18]. Nonlinear methods, such as the Takagi–Sugeno (TS) fuzzy model [19] and wavelet-based methods [20], have been shown superior to linear models.

Nowadays, the most common way to forecast the future values of a time series is the use of machine learning methods [21]. The reviewed literature shows that ANN methods have been successfully applied for forecasts of fluctuating energy supply.

These methods learn to recognize patterns in data using training datasets, and this is the main drawback: Historical data of both the weather forecast and the actual power production are needed to train the machine learning methods. Some studies showed that ANN models using multivariate, such as sun duration, temperature, wind speed, and relative humidity, can achieve much better performance than that using univariate [22].

Furthermore, the ANN methods are iterative procedures with a stochastic base: In fact, at the first iteration, weighted links among neurons are randomly set; then, they are optimized during iterations in order to minimize the error. For this reason, the resulting forecast depends on the specific trial (which is the result of a single forecast). Therefore, different trials can provide slightly different results within the same forecast. Usually the final trend of the forecast is the arithmetic average of the different trials led in a single run. This is called "ensemble method."

Any combination of two or more of the previously described methods is a hybrid model. The idea is to combine different models with unique features to overcome the single negative performance and finally improve the forecast. Recently, some papers show that all these methods need a phase of preprocessing the input datasets in order to increase the forecasting accuracy [23]. These models have been introduced to solve the weaknesses of individual methods and to enhance their strengths and accuracy. These combinations are called hybrid, blended, combined or ensemble methods. Hybrid models blend two or more techniques in various steps in order to deliver a single forecast. This combination can be obtained in two different ways: by blending two or more statistical techniques (hybrid-statistical) or from joining a statistical technique to a PV performance model (hybrid-physical).

Table 10.1 shows a possible timescale of RES energy forecasting. It includes very short-term, short-term, medium-term and long-term forecasting [24]. The forecast up

Table 10.1 Timescale classification for RES forecasting

Horizon	Range	Application
Very short	Few sec.–30'	Control and adjustment actions
Short	30'–6 h	Dispatch planning; load gain/drop
Medium	6 h–1 day	Generator on/off; operational security; electricity market
Long	1 day–1 week	Unit commitment; reserve requirement; maintenance schedule

to 24-h ahead or even more is needed for the power dispatching plans, the optimization operations of grid-connected RES plants and control of energy storage devices. Usually, the medium-term forecast is requested for the electricity market. The most common forecast horizon term for PV systems is 24-h ahead. Table 10.1 is set up with reference to wind power forecast, but it can be also applied with reference to PV. Anyway, forecasting term limits are not strictly defined, and some different specifications may be granted depending on the application of the forecasting model [15]. In this work, the "day-ahead" energy forecasting of multiple PV plants is considered, that is, the forecast of the hourly power produced in a day from a PV plant done 24 up to 48 h in advance.

10.3 Artificial Neural Networks

Artificial neural network is a machine learning technique that aims at mimicking the human brain and its neurons, through a fundamental unit, the perceptron, a concept introduced by Frank Rosenblatt in the period between early 1940 and late 1950 [25]. The structure of an artificial neural network [26] imitates those of the biological neural networks typical of the nervous system, composed of billions of neurons.

The human brain is a complex system, not linear and functioning as many processors in parallel, able to modify the connections among the neurons based on the acquired experience (i.e., able to learn). Another important characteristic of the brain is to be fault tolerant that means to continue functioning, even if a neuron (or some of its connections) is damaged, with degraded performance.

A good artificial model simulating the human brain should have such characteristics. A network made of simple elements is realized so as to be able to create a distributed system, highly parallel, capable to learn and able to generalize (i.e., to produce signals in output also in case of inputs never met before during the phase of the neural network training).

The artificial neuron, the basic unit of this network, has typically many inputs and only one output. Every input has a certain associated weight, which gives the conductivity of the correspondent input channel. The weighted total of the inputs of a neuron determines the neuron activation.

In order to train an artificial neural network, an input training set is supplied. Then, the output elaborated by the network for every input should be compared with the actual value. Evaluating the error (the difference between the two of them) weights are adjusted until the output given by the network produces an error which is under a certain fixed threshold.

10.3.1 Neuron Modeling

An artificial neuron, as shown in Fig. 10.2, is formed by *n* channels in input (called

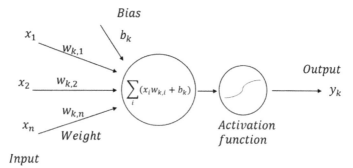

Fig. 10.2 Structure of an artificial neuron characterized by n inputs, weights, bias and by one output and a threshold

$x_1, x_2, \ldots, x_n$) each one having a weight W_i (which is a real number). The absolute value of a weight represents the connection force through the corresponding channel. If W_i is positive, we speak of "stirring up channel," while if W_i is negative, we speak of "inhibitory channel."

The output of the neuron is calculated applying the activation function (also called "transfer function") to the weighted amount of the inputs (also called "net").

The weighted amount of the inputs (called "a" for "activation level") is defined as follows:

$$a = \sum_{i=1}^{n} w_i \cdot x_i \tag{10.1}$$

Consequently, the output of a neuron is expressed as follows:

$$y = f(a) = f\left(\sum_{i=1}^{n} w_i \cdot x_i\right) \tag{10.2}$$

In the model shown in Fig. 10.2, we have also a bias having the effect of reducing the input value to the activation function, such bias (marked with b) influences the neuron output, which becomes:

$$y = f(a) = f\left(\sum_{i=1}^{n} w_i \cdot x_i - b\right) \tag{10.3}$$

The activation function gives back the output of an artificial neuron on the basis of the activation level supplied in input (generically marked with f in the equations). The output produced can be a real number belonging to a certain interval (i.e., [0, 1]) or an integer belonging to a discrete set (i.e., {0, 1} or {−1, 1}) depending on the activation function applied [27].

Among many activation functions, the sigmoid function (logistic function) has been adopted in this work:

$$y = f(a) = \frac{1}{1 + e^{-a}} \tag{10.4}$$

This (10.4) is one of the most utilized activation functions (together with the threshold functions) and can take on different shapes. Finally, continuous and differentiable activation functions are preferable for gradient-following algorithms. Activation functions spanning positive and negative values tend to expedite the rate of convergence, assuming a solution exists.

10.3.2 Error Indexes

In order to correctly define the accuracy of the prediction and the related error, it is necessary to define the indexes that can be used to evaluate the performances of the forecasting model. Some of these definitions come from statistics and are well known. In addition, technical papers present many of these indexes; therefore, here, we report some of the most commonly used error definitions. In this study, all the errors are referred only to the prediction in the daylight hours.

The normalized root mean square error NRMSE, based on the maximum observed power output $P_{m,h}$:

$$\text{nRMSE} = \frac{\sqrt{\frac{\sum_{h=1}^{N} |P_{m,h} - P_{p,h}|^2}{N}}}{\max(P_{m,h})} \cdot 100 \tag{10.5}$$

The weighted mean absolute error WMAE, based on total energy production:

$$\text{WMAE} = \frac{\sum_{h=1}^{N} |P_{m,h} - P_{p,h}|}{\sum_{h=1}^{N} P_{m,h}} \cdot 100 \tag{10.6}$$

In view of a more useful summary evaluation, additional performance metrics are proposed, aiming at providing a ranged value between 0 and 100% of the forecast accuracy.

For instance, the normalized mean absolute error NMAE, based on net capacity of the plant C:

$$\text{NMAE} = \frac{\sum_{h=1}^{N} |P_{m,h} - P_{p,h}|}{C} \cdot 100 \tag{10.7}$$

where N represents the number of samples (hours) considered: Usually, it is referred to a day, a month or a year. For this indicator, the rated power of the PV system was considered as C.

Furthermore, moving straight from NMAE definition, newer error metrics were proposed. The enveloped-weighted mean absolute error EMAE [28] is defined as:

$$\text{EMAE} = \frac{\sum_{h=1}^{N} |P_{m,h} - P_{p,h}|}{\sum_{h=1}^{N} \max(P_{m,h}, P_{p,h})} \cdot 100 \tag{10.8}$$

where the numerator is the sum of the absolute hourly errors, as in WMAE, while the denominator is the sum of the maximum between the forecast and the measured hourly power.

Finally, the objective mean absolute error (OMAE) [29] to take into account the variation of the theoretical maximum irradiation measurements throughout the year:

$$\text{OMAE} = \frac{\sum_{h=1}^{N} |P_{m,h} - P_{p,h}|}{\sum_{h=1}^{N} G_{\text{POA},h}^{\text{CS}}} \frac{G_{\text{STC}}}{C} \cdot 100 \tag{10.9}$$

where

- $G_{\text{POA},h}^{\text{CS}}$ is the solar irradiance on the plane of the array given by the clear sky solar irradiance model (CSRM) as it is described in [30, 31].
- G_{STC} is the solar irradiance at standard test conditions (equals 1000 W/m^2).

It is possible therefore to rewrite the OMAE highlighting the existing relationship with the former indicator NMAE:

$$\text{OMAE} = \text{NMAE} \frac{G_{\text{STC}}}{\sum_{h=1}^{N} G_{\text{POA},h}^{\text{CS}}} N \cdot 100 \tag{10.10}$$

It is worth highlighting that the forecast accuracies are not comparable site by site or hour by hour unless normalized. As a consequence, a benchmark is presented [32] and the skill score (SS) that measures the obtained improvement with respect to the persistence method [33] has the following formula:

$$\text{SS} = 1 - \frac{\text{nRMSE}}{\text{nRMSE}_{\text{pers}}} \tag{10.11}$$

Therefore for a perfect forecast SS is equal to 1; otherwise, a forecast output which is similar to the persistence forecast, that is, to consider the last measurements equal to the expected ones, will make SS resulting closer to 0.

As a general comment, it can be stated that NMAE is largely used to evaluate the accuracy of predictions and trend estimations. In fact, often relative errors are large because they are divided by small power values (for instance, the low values associated to sunset and sunrise): In such cases, WMAE could result very large and biased, while NMAE, by weighting these values with respect to the capacity C, is more useful. The NRMSE measures the average magnitude of the absolute hourly errors. In fact, it gives a relatively higher weight to larger errors, thus allowing to emphasize particularly undesirable results. Finally, it is noticeable how these final metrics are limited between 0 and 100%, providing an immediate indication on the magnitude of the daily diagnostic error.

10.3.3 Artificial Neural Network Setting Up

The main problem primarily consists in choosing how many neurons are in the layers and their possible layout in order to increase the forecast accuracy. Neurons in the input/output layers are fixed by the problem to be solved, while characterizing neurons in the hidden layers it is controversial, since the hidden neurons are regarded as the processing neurons in the network.

In addition, to have a small number of hidden layers might increase the speed of the training process; instead, a large number of hidden layers could extend it.

In order to properly characterize the number of neurons in the layers, there are different rules suggested in the literature from the tiling algorithm to the rule of thumb. More in general, many techniques are inspired by a "trial and error" basis consisting in a sensitivity analysis in order to achieve the minimum of the forecasting error by changing step by step a single parameter and observing the obtained result. Hence, there are two main symmetrical techniques from the already mentioned algorithm for selecting this parameter:

- "growing": the number of neurons in the hidden layers starts from a small number. Then it is gradually increased;
- "pruning": the number of neurons in the hidden layers starts from a large number, and then it is decreased by erasing the less significant components;

Usually, the initial number of hidden neurons h is selected in accordance with:

$$h = \frac{(i - o)}{2} \tag{10.12}$$

where i is the number of input and o is the number of output. Then, by choosing one of the previous methods, the best selection of hidden neurons could be set to achieve promising results for the network.

Besides theoretically speaking the nonlinearity of ANN increases with its size, which means simplifying that both the more complicated (nonlinear) the input/output relationship is, the more layers an ANN will need to model it, and that the greater the number of neurons per layer is, the more accurately ANN will be able to identify input/output relationships. However, for the most common applications, a single hidden layered ANN is able to accomplish the desired task as larger ANN are harder to train.

In order to perform the forecast, historical data are provided to the Neural Network Toolbox™ in MATLAB software [34], R2018a, MathWorks®, Natick, MA, USA, and grouped as: "training," "validation" and "test" dataset. Each group fulfills three specific tasks:

1. The *training set* includes samples which are employed to train the ANN. It should contain enough different examples (days) to make the network able to generalize its learning for future forecasts.
2. The *validation set* contains additional samples (i.e., days not already included in the training set) used by the network to check and validate the previous training process.
3. The *test set* is composed by the days forecasted by the network.

When dealing with the past measurements and historical weather forecast, it is reasonable to find a strategy in order to wisely assess and set up the forecasting method. In this work, a "roll-up moving window" approach has been adopted, as showed in Fig. 10.3. Namely, the training and validation set are made by the hourly samples belonging to the Nd days before the test set is provided (forecast day x_d) considering data as if they were in a temporal loop. Usually, the whole dataset consists of one year of data; therefore, it happens to employ samples in the training or validation, which are chronologically occurred after the test set. However, in order to keep the same amount of samples within the same training window size, this approach is preferable in order to avoid poorly trained networks for the first days of the dataset.

In this work, the moving window covers almost an entire year of available data, thanks to the implemented roll-up approach: The k days following the one selected

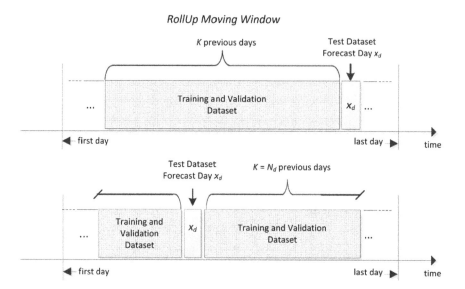

Fig. 10.3 Roll-up moving window technique

(forecast day x_d) are treated as belonging to the previous year, in order to obtain an entire available year. For example, if forecast day x_d is considered to be the 30th of June, the hourly samples belonging to the days starting from the 1st of July to the 31st of December are added before the days from the 1st of January to the 29th of June, in order to have a full year.

As it was shown here [28], it is worth to randomly shuffle the hourly samples belonging to the whole dataset each time the train and validation processes are performed and soon after to make an ensemble arithmetic average of the obtained output of the network. Therefore, in this work, after a sensitivity analysis [35], these ANN features were set:

- neurons in the first hidden layer: 11,
- neurons in the second hidden layer: 5,
- training and validation shares: 90 and 10% (excluding the test set),
- activation function: sigmoid,
- number of trials in the ensemble forecast: 40.

10.4 Physical Hybrid Artificial Neural Network

Hybrid models have been introduced to solve the weaknesses of individual methods and to enhance their strengths and accuracy. These combinations are called hybrid, blended or combined methods. Hybrid models mix two or more techniques in different steps of the process in order to deliver a single forecast. This combination can be obtained in two different ways: by blending two or more statistical techniques (hybrid-statistical) or from joining a statistical technique to a PV performance model (hybrid-physical). The ANN stochastic method here has been hybridized with a deterministic algorithm as the clear sky solar radiation model (CSRM) [31] in order to improve the overall performance. The CSRM for the next 24 h given as an additional input to the ANN together with the weather forecasts improves the sunrise and the sunset hours forecast and to exclude night time with null power output. Figure 10.4 shows the scheme of the Physical Hybrid Artificial Neural Network (PHANN) which has been adopted.

Historical data must be always validated, since unreliable data increase the odds of higher errors in the forecast. The preprocessing step initially includes the control of the coherence among the main variables measured in the PV plant, such as the solar radiation, the PV output power and the above-mentioned deterministic model of the solar radiation computed according to the geographical coordinates of the PV plant site.

Thus, the aim of using CSRM in this preliminary step is not only to determine the time span of the forecast between the sunrise and the sunset of each day, but also to validate the reliability of each sample. In this work, fifteen minutes data are available, and Fig. 10.5 shows the flowchart which has been used, as already explained in a previous work [36]. First, it starts acquiring the values of $G_{\mathrm{CRSM},1/4h}^{k}$, $G_{m,1/4h}^{k}$ and

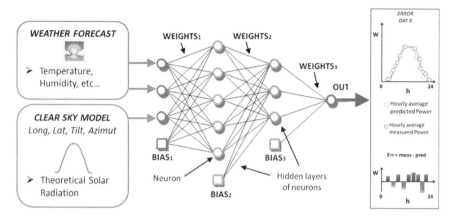

Fig. 10.4 PHANN scheme

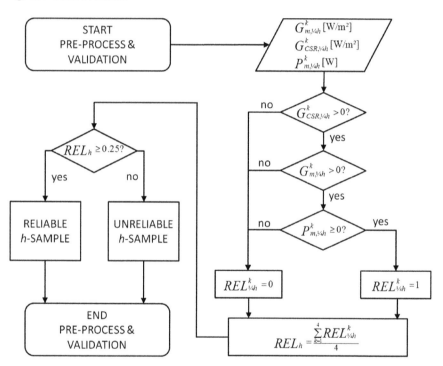

Fig. 10.5 Flowchart applied for data reliability employing clear sky algorithm

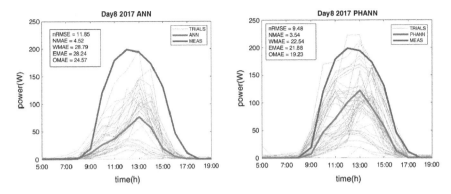

Fig. 10.6 Forecast performed by ANN (left) and by PHANN (right) in red. In gray, 40 trials of the ensembles are displayed and in blue, the actual measured power for day 8 of 2017

$P_{m,1/4h}^k$ which are the clear sky solar radiation, the measured solar radiation and the PV output power, respectively, in the kth quarter of an average hour sample.

Then, by comparing the other two variables, when the CSRM is positive, the reliability coefficient of the sample $REL_{1/4h}^k$ is equal to 1 if all the conditions occur at the same time; otherwise, it is equal to 0. When $G_{m,1/4h}^k$ is greater than 0 at the same time when $P_{m,1/4h}^k$ is equal to 0, the reliability coefficient of the sample $REL_{1/4h}^k$ is equal to 1.

Lastly, if REL_h, the hourly average of the four reliability coefficients in the same hour, is greater than 0, the hth hourly sample is considered in the next training and forecasting steps, otherwise not. Secondarily, there must be always correspondence between the number of samples and the time instant of the measured data and those provided by the Meteorological Service.

In the same paper [18], it is shown the improvement in the forecast after the PHANN for some significant days. For instance, in Fig. 10.6, a comparison between the day-ahead forecasts for the same day (01/15/2017) performed by the ANN (red line on the left) and by the PHANN (red line on the right) is provided. As it is possible to see, by comparing the relevant error indicators, PHANN forecast was more accurate than the purely stochastic ANN (without CSRM).

Deterministic forecasting models usually are based on a few atmospheric parameters. Therefore, they strongly depend on the accuracy of the data provided by the weather forecast services. Instead, stochastic methods are more flexible because they are able to infer relationships existing among different parameters affected by inaccuracy and the actual variable to forecast.

A meaningful example of the weather forecast accuracy is shown in Fig. 10.7. Here (on the left), for a single PV module, the daily power forecast performed by two deterministic algorithms on the basis of the weather forecast given 72 h in advance is displayed. On the right, the actual power of 29 PV modules is depicted, and it is clear that the weather forecast was highly inaccurate for that day (03/10/2017). Weather forecasts provided much closer to the day of interest are more accurate. On the other

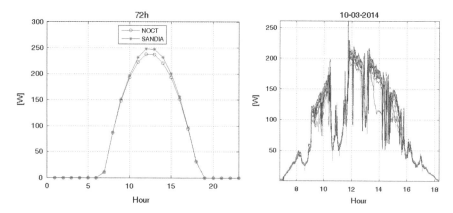

Fig. 10.7 (Left) Daily power forecast for one PV module performed by two deterministic algorithms based on the weather forecast for 03/10/2017 given 72 h in advance. (Right) The actual power values during that day of 29 PV modules

hand, in order to join the day-ahead electricity market, there is a hard time limit to perform the forecast.

We employ an hybrid tool PHANN, combining deterministic algorithm (CSRM) and stochastic forecasting method (ANN), to partly overcome the inaccuracy of the weather forecast, as it will be shown later, which is at the moment an unavoidable barrier in the day-ahead PV plant power forecast.

10.5   Case Study

The data on which is based the first part of the present work have been collected at the SolarTech Lab, in the energy department at Politecnico di Milano, located at latitude 45.502941° N and longitude 9.156577° E. The laboratory is equipped with 29 modules, of different typology: Ten of them are made by Si-monocrystalline, eleven Si-polycrystalline and five hybrid technology. All the modules have a tilt angle of 30° (0° representing a horizontal module parallel to the ground) and an azimuth angle of $-6° 30'$, assuming 0° being the south direction and positive angles moving westward. Each PV module is equipped with a micro-converter connecting it to the electric grid, allowing them to work independently and generally speaking always at their maximum power point.

A meteorological private company provides the weather forecast, while the actual environmental conditions are monitored through the meteorological station which is equipped with: solar irradiation sensors (a DNI sensor and two pyranometers measuring the total and diffuse irradiance on the horizontal plane), temperature-humidity sensors, wind speed and direction sensors and rain collector [37]. The dataset with the forecast meteorological quantities was composed by the following information:

- Dry bulb air temperature—T_{amb} (°C),
- Global horizontal irradiation—GHI (W/m^2),
- Global irradiation on the plane of the array—G_{POA} (W/m^2),
- Wind speed—Ws (m/s),
- Wind direction—Wd (°),
- Atmospheric pressure—AP (hPa),
- Precipitation—PR (mm),
- Cloud cover—CC (%),
- Cloud type—CT (low/medium/high).

Together with these forecast quantities, the measured DC power production PDC [W] for the corresponding time interval is available. The dataset refers to the monocrystalline PV module Aleo Solar 285 W and to the year 2017, displayed in Fig. 10.8. It has been at first cleaned from the days with faulty or missing measurements, resulting in 269 available days in 2017 and 225 available days in 2018.

The previously described Physical Hybrid Artificial Neural Network (PHANN) has been trained on these datasets. First, the network has been trained on the measured data, to assess the performance of the network and the relative importance of the meteorological error on the final forecast. Specifically, the input dataset was composed by the following information of recorded parameters:

- Dry bulb air temperature—T_{amb} (°C),
- Global horizontal irradiation—GHI (W/m^2),
- Global irradiation on the plane of the array—G_{POA} (W/m^2).

Fig. 10.8 Panoramic view of the Solar Tech Lab. The Aleo Solar PV module is highlighted

After, ANN has been trained on the meteorological forecast related to the year 2017, running 40 parallel trials in order to be able to perform a subsequent ensemble analysis, to lower the stochasticity of the results.

10.6   Results and Discussion

First, it is reasonable to wonder how much the weather forecasts inaccuracy could affect the PV output power forecast of the chosen method. In order to estimate this error, the choice to include the best possible weather forecasts was adopted. In this case, the accuracy of the weather forecasts employed in the current analysis by means of the error metrics when applicable. As the measurements available were the dry bulb ambient temperature (T_{amb}) and the solar irradiation on the plane of the array (G_{POA}), these two parameters were compared with the values in the weather forecasts for the year 2017. OMAE was obviously not applicable in the T_{amb} forecast as it deals with the solar irradiation values. And likewise with NMAE, the net capacity of the plant C expressed in the former Eq. (10.7) would be meaningless. As a consequence it has been changed as NMAE* where the MAE has been normalized by the maximum measured given parameter (Xm,h): NMAE* = MAE/max(Xm,h)·100

$$\text{NMAE}^* = \frac{\sum_{h=1}^{N} \left| P_{m,h} - P_{p,h} \right|}{\max\left(P_{m,h} \right)} \cdot 100 \qquad (10.13)$$

The results of this preliminary analysis are summed in Table 10.2. As it can be seen the weather forecast inaccuracy of the parameters employed in the forecast process will affect the output of the forecasting method differently.

As deterministic methods are directly connecting weather parameters to the desired power output according to a physical model of the PV system, it is reasonable to infer that the inaccuracy of the PV power forecasted by means of this class of forecasting models will be in the same order of magnitude of the inaccuracy of the weather forecasts provided as input to the physical model.

On the other hand, machine learning techniques have the strength and flexibility of inferring the existing links among the above-mentioned parameters by acquiring them from data. In our case by analyzing the PV power output forecast provided by PHANN, which has been trained with the actual measured values in the PV plant, we obtain the results which are summed in Table 10.3. It is noticeable how all the error metrics are considerably lowered if compared to the previous ones.

Table 10.2   Weather forecasts accuracy versus actual measurements, year 2017

	nRMSE	WMAE	NMAE*	EMAE	OMAE
T_{amb}	13.4	19.57	4.04	14.67	n.a.
G_{POA}	35.44	89.15	4.81	28.64	18.57

Table 10.3 Performance of the PHANN forecast trained with actual measurements, year 2017

	NRMSE	WMAE	NMAE	EMAE	OMAE
PHANN(m)	2.22	5.08	0.58	4.68	2.09

We can presumably infer that PHANN learned the inaccuracies in the weather forecasts by coupling them, in addition to the CSRM, to the actual power of the PV plant. This kind of comparison has been deeply assessed in the previous works [38, 39]. The here obtained results highlight how much the meteorological forecast error affects the final prediction of the power production.

Moving forward in the current analysis, it is possible to compare the error metrics related to the forecast obtained by means of three different methods: the persistence, the purely stochastic ANN and the hybrid PHANN.

The forecast has been performed on the available days, for the year 2017 and 2018 with the roll-up moving window techniques and they are assessed, with the metrics previously exposed, which are shown in Table 10.4.

Persistence method, which is commonly adopted as the benchmark, committed considerably the highest error which has been calculated. This is easy to understand as we are dealing with the day-ahead forecast, and weather conditions might be strongly different from one day to the following with more than 24 h in advance. Consequently, this kind of method is preferable for the short- or very short-term forecast rather than the day-ahead forecast.

Here, we can also notice which PHANN trained with the measured (Table 10.3) data shows a NMAE significantly lower (82.3%) with respect to the one trained on the forecast meteorological data, pointing out the importance of an accurate weather forecast provided as input. The reported additional error metrics are in agreement with what has been stated above.

The average improvement scored by PHANN against ANN on both of the years is related to a general trend which can be seen in Fig. 10.9. Here, on the left, it is noticeable how generally PHANN is better forecasting in more days of 2018 than ANN. On the right, the one of best daily PHANN forecast is reported. The same trend is confirmed in the year 2017.

Table 10.4 Different forecasting methods comparison in different years

Year	Method	nRMSE	WMAE	NMAE	EMAE	OMAE	SS
2017	Persistence	34.26	84.35	5.61	32.47	21.65	0
	ANN	18.75	46.31	4.17	27.19	15.48	0.45
	PHANN	18.07	43.73	3.96	26.09	14.79	0.47
2018	Persistence	41.47	113.53	5.34	39.15	20.32	0
	ANN	26.32	62.26	3.77	32.35	14.37	0.36
	PHANN	26.07	61.89	3.67	31.83	14.1	0.37

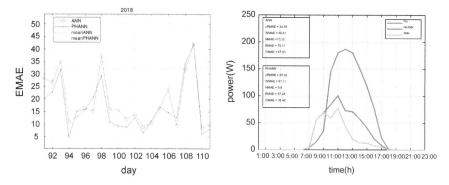

Fig. 10.9 Daily EMAE comparison (on the left), ANN in blue and PHANN in red, the relevant mean values are also added. One day comparison (on the right) where PHANN in red is scoring better than ANN in yellow, with respect to the actual measured power P_m in blue

10.7 Conclusion

This chapter was focused on the arrangement of a hybrid method (PHANN—Physic Hybrid Artificial Neural Network) combining the stochastic ANN method as well as the deterministic clear sky solar radiation model (CSRM).

The setting up of this method, which is repeatable for any PV plant, allowed to identify the best settings in terms of: number of neurons, neurons layout and training set size in order to perform the day-ahead power production forecast with a particular attention to the assessment of the forecasting accuracy.

This method was validated with data recordings from a real PV plant in Milan, Italy, for two consecutive years (2017 and 2018). Concerning the setting up of the method and the day-ahead forecast accuracy, it can generally be inferred that the here presented hybrid method (PHANN) better performs the day-ahead forecast if compared to pure stochastic models (ANN) and the persistence. Nevertheless, these forecasts are strongly affected by the weather historical data accuracy used to train them; therefore, it is possible that a properly trained network could still provide an inaccurate profile due to the weather forecasts inaccuracy.

References

1. Mitcell, R.B.: International Environmental Agreements Database Project (Version 2014.3) [Online]. Available: https://iea.uoregon.edu/. Accessed 20 Dec 2019
2. OECD/IEA and IRENA: Perspectives for the Energy Transition: Investment Needs for a Low-Carbon Energy System, p. 204. International Energy Agency (2017)
3. IEA: Executive Summary of the World Energy Investment, 2018. International Energy Agency (IEA) [Online]. Available: https://webstore.iea.org/download/summary/1242?fileName=English-WEI-2018-ES.pdf (2018). Accessed 20 Dec 2019

4. Renewables 2018. Solar Energy [Online]. Available: https://www.iea.org/topics/renewables/solar/. Accessed 20 Dec 2019
5. Coimbra, C.F.M., Kleissl, J., Marquez, R.: Overview of solar-forecasting methods and a metric for accuracy evaluation. In: Solar Energy Forecasting and Resource Assessment, pp. 171–193 (2013)
6. World Meteorological Organization: Manual on the Global Data-Processing and Forecasting System, vol. WMO-No. 48, no. 485 (2010)
7. Das, U.K., et al.: Forecasting of photovoltaic power generation and model optimization: a review. Renew. Sustain. Energy Rev. **81**, 912–928 (2018)
8. Ulbricht, R., Fischer, U., Lehner, W., Donker, H.: First steps towards a systematical optimized strategy for solar energy supply forecasting. In: ECML/PKDD 2013, 1st International Workshop on Data Analytics for Renewable Energy Integration, pp. 14–25 (2013)
9. Mellit, A., Pavan, A.M.: A 24-h forecast of solar irradiance using artificial neural network: application for performance prediction of a grid-connected PV plant at Trieste, Italy. Sol. Energy **84**(5), 807–821 (2010)
10. Reikard, G.: Predicting solar radiation at high resolutions: a comparison of time series forecasts. Sol. Energy **83**(3), 342–349 (2009)
11. Izgi, E., Öztopal, A., Yerli, B., Kaymak, M.K., Şahin, A.D.: Short-mid-term solar power prediction by using artificial neural networks. Sol. Energy **86**(2), 725–733 (2012)
12. Shi, J., Lee, W.J., Liu, Y., Yang, Y., Wang, P.: Forecasting power output of photovoltaic systems based on weather classification and support vector machines. IEEE Trans. Ind. Appl. **48**(3), 1064–1069 (2012)
13. Pedro, H.T.C., Coimbra, C.F.M.: Assessment of forecasting techniques for solar power production with no exogenous inputs. Sol. Energy **86**(7), 2017–2028 (2012)
14. Monteiro, C., Fernandez-Jimenez, A., Ramirez-Rosadoc, I., Munoz-Jimenez, A., Lara-Santillan, P.: Short-term forecasting models for photovoltaic plants: analytical versus soft-computing techniques. Math. Probl. Eng. 9 (2013)
15. Wang, F., Mi, Z., Su, S., Zhao, H.: Short-term solar irradiance forecasting model based on artificial neural network using statistical feature parameters. Energies **5**(5), 1355–1370 (2012)
16. Yang, H.-T., Chao-Ming, H., Huang, Y.-C., Yi-Shiang, P.: A weather-based hybrid method for one-day ahead hourly forecasting of PV power output. In: Proceedings of 2014 9th IEEE Conference on Industrial Electronics and Applications, ICIEA 2014, vol. 5, no. 3, pp. 526–531 (2014)
17. Dolara, A., Lazaroiu, G.C., Leva, S., Manzolini, G.: Experimental investigation of partial shading scenarios on PV (photovoltaic) modules. Energy **55**, 466–475 (2013)
18. Bacher, P., Madsen, H., Nielsen, H.A.: Online short-term solar power forecasting. Sol. Energy **83**(10), 1772–1783 (2009)
19. Iqdour, R., Zeroual, A.: A rule based fuzzy model for the prediction of daily solar radiation. In: Proceedings of IEEE International Conference on Industrial Technology, vol. 3, pp. 1482–1487 (2004)
20. Capizzi, G., Bonanno, F., Napoli, C.: A wavelet based prediction of wind and solar energy for long-term simulation of integrated generation systems. In: SPEEDAM 2010—International Symposium on Power Electronics, Electrical Drives, Automation and Motion, no. 3, pp. 586–592 (2010)
21. Chen, S.H., Jakeman, A.J., Norton, J.P.: Artificial intelligence techniques: an introduction to their use for modelling environmental systems. Math. Comput. Simul. **78**(2–3), 379–400 (2008)
22. Sfetsos, A., Coonick, A.H.: Univariate and multivariate forecasting of hourly solar radiation with artificial intelligence techniques. Sol. Energy **68**(2), 169–178 (2000)
23. Ogliari, E., Grimaccia, F., Leva, S., Mussetta, M.: Hybrid predictive models for accurate forecasting in PV systems. Energies **6**(4), 1918–1929 (2013)
24. Soman, S.S., Zareipour, H., Malik, O., Mandal, P.: A review of wind power and wind speed forecasting methods with different time horizons. In: North American Power Symposium 2010, NAPS 2010, pp. 1–8 (2010)

25. Rosenblatt, F.: Principles of neurodynamics: perceptron and the theory of brain mechanism. Am. Math. Mon.VG-1196-G-8, 3–621 (1961)
26. Krose, B., van der Smagt, P.: Introduction to neural networks. Int. J. Join. Mater. **6**(1), 4–6 (1994)
27. Jain, A.K., Mao, J., Mohiuddin, K.M.: Artificial neural networks: a tutorial. Computer (Long. Beach. Calif.) **29**(3), 31–44 (1996)
28. Dolara, A., Grimaccia, F., Leva, S., Mussetta, M., Ogliari, E.: Comparison of training approaches for photovoltaic forecasts by means of machine learning. Appl. Sci. **8**(2), 228 (2018)
29. Leva, S., Mussetta, M., Ogliari, E.: PV module fault diagnosis based on micro-converters and day-ahead forecast. IEEE Trans. Ind. Electron. 1 (2018)
30. Kasten, F., Czeplak, G.: Solar and terrestrial radiation dependent on the amount and type of cloud. Sol. Energy **24**(2), 177–189 (1980)
31. Bird, R.E., Riordan, C.: Simple solar spectral model for direct and diffuse irradiance on horizontal and tilted planes at the earth's surface for cloudless atmospheres. J. Clim. Appl. Meteorol. **25**, 87–97 (1986)
32. Coimbra, C.F.M., Kleissl, J., Marquez, R.: Chapter 8—overview of solar-forecasting methods and a metric for accuracy evaluation. In: Kleissl, J. (ed.) Solar Energy Forecasting and Resource Assessment, pp. 171–194. Academic Press, Boston (2013)
33. Murphy, A.H.: Skill scores based on the mean square error and their relationships to the correlation coefficient. Mon. Weather Rev. **116**, 2417–2424 (1988)
34. Beale, M.H., Hagan, M.T., Demuth, H.B.: Neural Network Toolbox™ User's Guide. Math-Works Inc, USA (1992)
35. Grimaccia, F., Leva, S., Mussetta, M., Ogliari, E.: ANN sizing procedure for the day-ahead output power forecast of a PV plant. Appl. Sci. **7**(6) (2017)
36. Leva, S., Dolara, A., Grimaccia, F., Mussetta, M., Ogliari, E.: Analysis and validation of 24 hours ahead neural network forecasting of photovoltaic output power. Math. Comput. Simul. **131**, 88–100 (2017)
37. Leva, S., Mussetta, M., Nespoli, A., Ogliari, E.: PV power forecasting improvement by means of a selective ensemble approach. In: 2019 IEEE Milan PowerTech, pp. 1–5 (2019)
38. Ogliari, E., Dolara, A., Manzolini, G., Leva, S.: Physical and hybrid methods comparison for the day ahead PV output power forecast. Renew. Energy **113**, 11–21 (2017)
39. Ogliari, E., Niccolai, A., Leva, S., Zich, R.E.: Computational intelligence techniques applied to the day ahead PV output power forecast: PHANN, SNO and mixed. Energies **11**(6), 1487 (2018)

Chapter 11
XSG-Based Control of an Autonomous Power System

N. Chettibi and A. Mellit

Abstract This chapter deals with the control strategy for a stand-alone photovoltaic/wind/battery energy generation system. The goal is to examine the performance of the power system under several conditions of generation and demand. A centralized power management system is established to supervise the power flow between the generation units and user loads. Local controllers of the renewable sources and storage system are designed based on classical control methods. The overall system is simulated in the MATLAB/Simulink environment, using Xilinx System Generator (XSG) tool, for possible implementation of generated VHDL (Hardware Description Language) code on a reconfigurable Field-Programmable Gate Array (FPGA) card. The simulation results are provided in order to demonstrate the accuracy and feasibility of the designed control scheme.

Keywords Photovoltaic · Wind turbine · Battery · Load demand · Power management · Control · Xilinx system generator

11.1 Introduction

In the last decades, much more interest is given for Hybrid Energy Generation Systems (HEGS) designed for different applications (water pumping, residential application, grid-connected systems, etc.). The HEGS based on Renewable Energy Sources (RES) (photovoltaic, wind, hydraulic, etc.) looks very promising since they can produce a clean energy without pollutants emissions. In stand-alone HEGS, the use of energy storage devices (like batteries, ultra-capacitors, flywheel, etc.) is of main interest in order to ensure an interruptible power supply of end-user's loads (of type AC and/or DC) [1].

The intermittent nature of RESs (solar and wind powers) together with irregular variation of load demand leads to power unbalance between the generation units and

N. Chettibi (✉) · A. Mellit
Renewable Energy Laboratory, University of Jijel, Jijel, Algeria
e-mail: chettibi.na@gmail.com

© The Editor(s) (if applicable) and The Author(s), under exclusive license
to Springer Nature Switzerland AG 2020
A. Mellit and M. Benghanem (eds.), *A Practical Guide for Advanced Methods in Solar Photovoltaic Systems*, Advanced Structured Materials 128,
https://doi.org/10.1007/978-3-030-43473-1_11

customers. In order to ensure a coordinated control of the power flow among HEGS components, a Power Management System (PMS) should be carefully designed. Its principal goals are to ensure continuous satisfaction of consumers' demand, to stabilize the renewable power generation with the highest efficiency and to extend the storage's lifetime [1, 2]. It gives the appropriate power set-points for local controllers of power generation units [2].

A large number of research works have been done in the topic of control of HEGS operating in stand-alone mode or grid-connected mode. Ref. [3] presents an energy management system for a water pumping system making use of PV and wind power sources with battery storage bank. A study of a HEGS composed of a wind turbine, a solar array, a microturbine and a battery storage system is performed in [4]. The authors have applied several control approaches like feedback linearization method and fuzzy logic in addition to a centralized PMS that supervises the power flow in the multisource system. Three power management algorithms are proposed in [5] in order to ensure the autonomy of hybrid PV/wind/battery system applied for water pumping. The management strategy of a grid-connected hybrid wind, photovoltaic and flywheel energy storage system for residential applications is presented in Ref. [6]. In [7], a PMS and a predictive control scheme are proposed for a hybrid AC/DC microgrid. A hardware in the loop simulation is performed to demonstrate the effectiveness of the presented control strategy. In Ref. [2], a validation of a two-level energy management strategy proposed for a PV-fuel cell-battery-based DC microgrid is performed. *Sharma* et al. in [8] have proposed and validated experimentally a dynamic PMS for autonomous hybrid PV-FC-battery-supercapacitor microgrid. In Ref. [9], a fuzzy logic-based PMS is suggested for the supervision of a stand-alone PV-FC-electrolyzer-battery-based hybrid system. An investigation on the control of autonomous microgrid is performed in [10] with an appropriate strategy for the power flow management. Based on RT-LAB real time simulation platform, the authors in [11] have performed a hardware in the loop simulation of a PMS designed for a DC microgrid.

The main goal of this chapter is to design and control an autonomous hybrid PV/wind/battery power system. Hence, simple local controllers are designed to harvest the maximum power from the renewable energy sources as well as to manage the operating mode of the battery storage system. A central PMS is developed to ensure the proper operation of the stand-alone system with variable generation and load conditions. The control system is constructed by the means of XSG tool that allows the generation of VHDL code. The goal is to realize an experimental set-up of the hybrid system based on a FPGA platform.

11.2 Architecture and Control of the HEGS

The structure of the hybrid power system under study is shown in Fig. 11.1. It consists of four Wind Turbine Generators (WTG) (For total power of 2.4 kW), three photovoltaic strings mounted in parallel (for nominal power more than 2.24 kW)

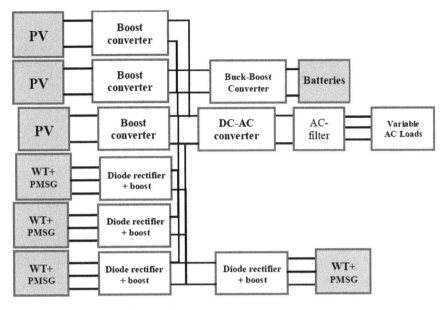

Fig. 11.1 Configuration of the hybrid power system under study

and a bank of lithium-ion batteries. The PV Generator (PVG) is interfaced with
the DC bus through a boost converter. Each WTG is composed of a vertical axis
wind turbine (WT) feeding a Permanent Magnetic Synchronous Generator (PMSG),
which is interconnected to the DC link through a diode rectifier and a controlled
boost converter.

11.2.1   Control of the PV Array

The well-known single diode model [1] is adopted in this study for the modelling
of the multi-string PV array (see Fig. 11.2). It consists of N_P strings connected in
parallel where, each one is composed of N_S modules connected in series. The output
current of each PV string can be calculated according to [1]:

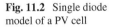

Fig. 11.2 Single diode
model of a PV cell

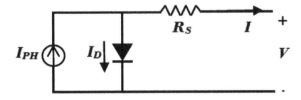

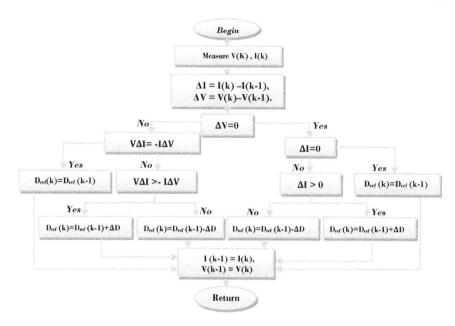

Fig. 11.3 Adopted IncCond algorithm [12]

$$I = I_{PH} - I_S\left[\exp\left(\frac{q}{aKT}\left(\frac{V}{N_S} + IR_S\right)\right) - 1\right] \qquad (11.1)$$

I and V are the PV string current and voltage, I_{PH} is the photocurrent, I_S is the saturation current, q is the electron charge, K is the Boltzmann constant, a is the diode ideality factor and T is the temperature. R_S is the series resistance.

In order to ensure an optimal operation of the PVG under varying climatic conditions, a maximum power point tracking (MPPT) algorithm has to be implemented. In this work, the incremental conductance (IncCond) method [12] is chosen to accomplish the MPPT task (see Fig. 11.3). The IncCond gives at the output the correct value of the duty cycle for the boost converter that interfaces the PVG. The block scheme of the XSG-based IncCond controller is illustrated in Fig. 11.4. Our design goal here is to minimize the FPGA's resources consumption by the IncCond controller by avoiding the implementation of division operations.

11.2.2 Control of the Wind Power System

The aerodynamic power generated by the vertical axis wind turbine can be calculated according to the following formula [13]:

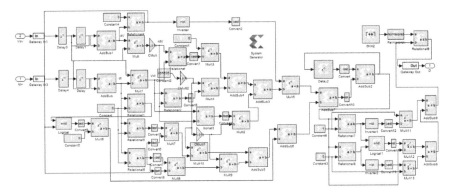

Fig. 11.4 XSG-based control circuit of boost converter

$$P_t = \frac{1}{2}\rho SC_P V_W^3 = \frac{1}{2}\rho RHC_P V_W^3 \tag{11.2}$$

where ρ, V_W, R and H are the air density, the wind speed, the radius and height of the turbine, respectively. The power coefficient C_P [13]:

$$C_P(\lambda) = -0.2121\lambda^3 + 0.0856\lambda^2 + 0.2539\lambda$$

λ is the tip speed ratio. The electromagnetic torque T_e of the PMSG is expressed as [13]:

$$T_e = \frac{3}{2}p\big((L_{sd} - L_{sq})I_{sq}I_{sd} + \phi I_{sq}\big) \tag{11.3}$$

Such that: p is the number of pole pairs; Φ is the magnetic flux; L_{sd}, L_{sq}, I_{sd}, I_{sq} are, respectively, the d-axis and q-axis inductances and currents of the PMSG. The MPPT control of the wind turbine generator is performed using the perturbation and observation (P&O) algorithm [14]. The control scheme adopted for the control of the boost converter interfacing the diode rectifier is presented in Fig. 11.5. The P&O algorithm increments or decrements the set-point of the boost converter's inductance current (I_L) based on the change in the mechanical power and rotational speed as presented in Fig. 11.6.

11.2.3 Control of the Battery Storage System

A bidirectional buck-boost converter is used to interface the battery bank with the DC link. The charge/discharge process of the storage system is supervised by regulating the output power to its reference provided by the PMS. A PI regulator is so used to minimize the power error and to determine the correct duty ratio of the Pulse Width

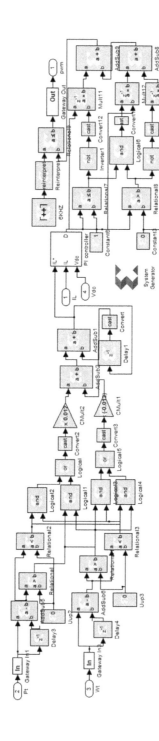

Fig. 11.5 XSG-based control scheme of boost converter interfacing the WTG

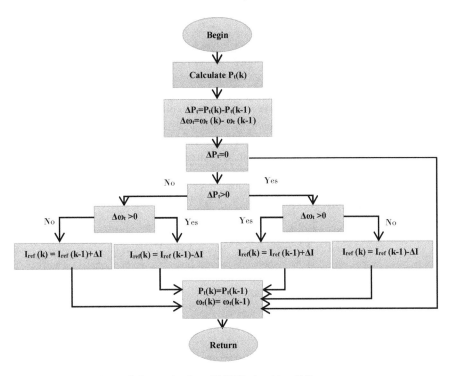

Fig. 11.6 Perturbation and observation-based MPPT algorithm [14]

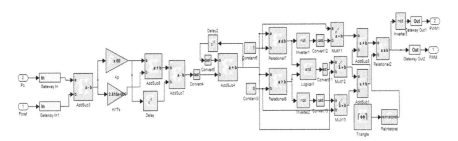

Fig. 11.7 Control scheme of the bidirectional converter interfacing the battery bank

Modulation (PWM) signals applied to the switching devices. The adopted control scheme is illustrated in Fig. 11.7.

11.2.4 Control of the Inverter

The three-phase inverter has to supply the power at the load side with the same magnitude and frequency of the AC loads voltage [15]. For this goal, the control scheme presented in Ref. [15] is adopted for the control of the autonomous inverter as presented in Fig. 11.8. The vector of the inverter output voltage is regulated in the rotating frame (dq) through PI regulators. The d-axis and q-axis current references $(i_d^*$ and $i_q^*)$ are generated at the output of the d-axis and q-axis voltage controllers [15]. The inner current control loop is established based on PI controllers with elimination of d- and q-axis coupling effect [15].

The d-axis and q-axis components of the load voltage $(v_d$ and $v_q)$ can be expressed in function of the current components $(i_d$ and $i_q)$ as follows:

$$\begin{cases} v_d = v_{id} - Ri_d - L\frac{di_d}{dt} + \omega Li_q \\ v_q = v_{iq} - Ri_q - L\frac{di_q}{dt} - \omega Li_d \end{cases} \tag{11.4}$$

R, L are the resistance and inductance of the first-order filter. v_{id} and v_{iq} are the d-axis and q-axis components of the inverter output voltage. The d-axis and q-axis references of the inverter voltage have to be calculated as:

$$\begin{cases} v_{id} = v_d + (K_P \cdot \Delta i_d + K_I \int \Delta i_d dt) - \omega Li_q \\ v_{iq} = v_q + (K_P \cdot \Delta i_q + K_I \int \Delta i_q dt) + \omega Li_d \end{cases} \tag{11.5}$$

where K_P and K_I are the proportional and integral gains of PI controller. Δi_d and Δi_q are, respectively, the instantaneous errors of the d-axis and q-axis currents:

$$\begin{cases} \Delta i_d = i_d^* - i_d \\ \Delta i_q = i_q^* - i_q \end{cases} \tag{11.6}$$

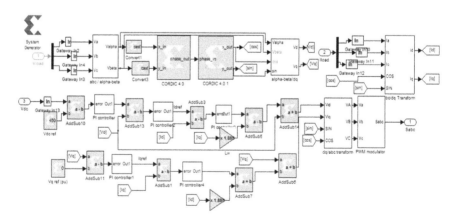

Fig. 11.8 XSG-based control scheme of load-side inverter

The voltage components (v_{id}, v_{iq}) are then transformed into the natural frame (abc) in order to be used by the PWM block. The details of the inverter control scheme developed using the XSG tool are given in Fig. 11.8.

11.2.5 Power Management System

According to the values of the power demand, the generated power and the state of charge (SOC) of battery, the central PMS decides on the value of the power reference (P_{bref}) of the storage system. The minimum and maximum values of the SOC of the battery bank are defined as: $SOC_{min} = 20\%$ and $SOC_{max} = 80\%$. The net power (P_{net}) is calculated as the difference between the load power and the power generated by renewable PV and wind power generators: $P_{net} = P_{Load} - (P_{PV} + P_{WT})$.

In this study, the AC loads are categorized into two types: critical and non-critical loads. The critical loads must still connected regardless the generation and demand conditions. In the case where $SOC < SOC_{min}$ and the renewable generators cannot meet the load demand, the non-critical loads can be disconnected in order to guarantee the continuous supply of critical loads. The operating mode of the storage is so defined according to the following supervision rules:

Rules 1: *if* $SOC < SOC_{min}$ *and* $P_{net} = 0$ *then charge batteries bank.*
Rules 2: *if* $SOC > SOC_{min}$ *and* $P_{net} = 0$ *then idle mode of batteries bank.*
Rules 3: *if* $SOC < SOC_{max}$ *and* $P_{net} < 0$ *then charge batteries bank.*
Rules 4: *if* $SOC > SOC_{max}$ *and* $P_{net} < 0$ *then idle mode of batteries bank.*
Rules 5: *if* $SOC < SOC_{min}$ *and* $P_{net} > 0$ *then charge batteries bank.*
Rules 6: *if* $SOC > SOC_{min}$ *and* $P_{net} > 0$ *then discharge batteries bank.*

More details on the adopted PMS are given in Fig. 11.9. Also, Fig. 11.10 illustrates the PMS designed using the XSG blocksets.

11.3 Simulation Results

In order to validate the performance of the HEGS under study, two simulations are performed for two scenarios. The system parameters are given in Table 11.1.

Case 1: Variable load, constant generation and $SOC < SOC_{min}$:

In this scenario, the HEGS is studied for stable climatic conditions (STC conditions and constant nominal wind speed $V_W = 15$ m/s). The initial state of charge of the battery bank is considered 15%, and the load demand varies as described in Fig. 11.13.

In this case, the RESs operate at their optimum power points that correspond to the nominal climatic conditions. As can be seen from Fig. 11.13, the RESs generate sufficient power to charge battery bank and to supply the AC loads. The PMS changes

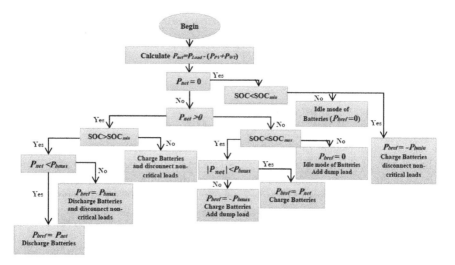

Fig. 11.9 Adopted PMS algorithm

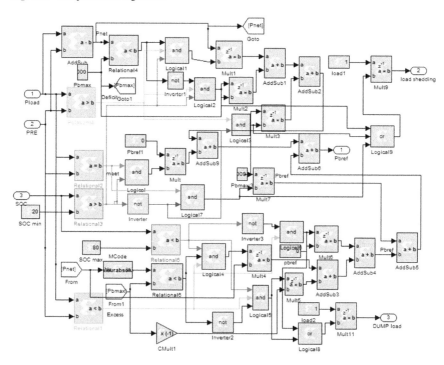

Fig. 11.10 Implemented PMS based on XSG tool

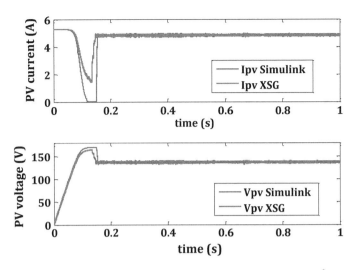

Fig. 11.11 Simulation results of the XSG-IncCond controller for $G = 1000$ W/m^2 and $T = 25$ °C

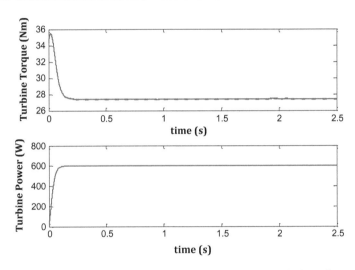

Fig. 11.12 Simulation results of the XSG-WTG controller for nominal wind speed $V_W = 15$ m/s

Table 11.1 Simulation parameters

PV module MSX-83	Value	WTG parameter	Value
Maximum power (P_{max})	83 W	Height of turbine	2 m
Voltage at maximum power (V_{MPP})	17.1 V	Radius	0.5 m
Current at maximum power (I_{MPP})	4.85 A	Stator resistance	1.137 Ω
Short-circuit current (I_{sc})	5.27 A	Stator inductance	2.7mH
Open-circuit voltage (V_{oc})	21.2 V	Magnetic flux	0.14 Wb
Number of cells in series	36	Pole pairs	17

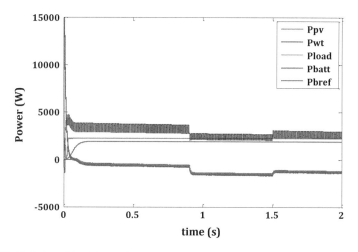

Fig. 11.13 Variation of load demand, power generated by RESs and storage system in the first test

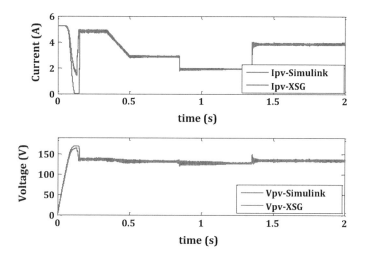

Fig. 11.14 Simulation results of the XSG-IncCond controller under varying irradiance

the value of the power set-point of the battery bank in order to work it continuously in the charge mode. On the other hand, it can be seen from Figs. 11.11 and 11.12 that the XSG-based controllers perform well with high precision and a big similarity with the results obtained using the classical Simulink-based controllers. The quantization errors are acceptable but surely can be enhanced.

Case 2: Variable load, variable generation and $SOC_{min} < SOC < SOC_{max}$:

In the second simulation case, the HEGS undergoes to a variable irradiation level and random wind speed. The load demand varies as shown in Fig. 11.16, whereas

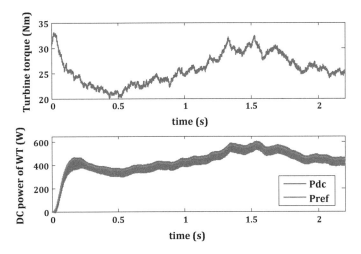

Fig. 11.15 Simulation results of the XSG-WTG controller for varying wind speed

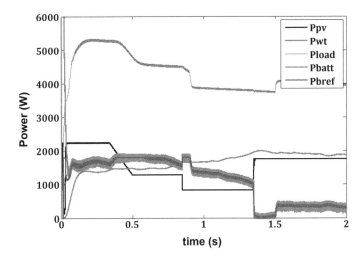

Fig. 11.16 Variation of load demand, power generated by RESs and storage system in the second test case

the initial SOC of battery bank is considered 56%. It is obvious from Figs. 11.14 and 11.15 that the local controllers of the PVG and WTG perform wells and ensure continuous tracking of the MPPs irrespective of meteorological conditions change. The XSG-based controllers give acceptable results with high accuracy and good tracking performance. A report on the total consumption of the FPGA's resources by the developed controllers is given in Table 11.2. On the other side, with reference to Fig. 11.16, it can be observed that the PMS works correctly and gives for each step

Table 11.2 VIRTEX 6 FPGA resource utilization

FPGA component	Controller			
	PVG	Battery	PMS	Inverter
Slice LUTs	0% (238)	0% (30)	0% (153)	0% (2354)
Slice registers	0% (155)	0% (75)	0% (1)	0% (1915)
Bounded IOBs	5% (66)	6% (75)	12% (151)	23% (283)
BUFG/BUFGCTRLs	3% (1)	3% (1)	3% (1)	3% (1)
LUTs-FF pairs	1% (4)	1% (2)	0% (0)	29% (970)
DSP48E1s	–	–	–	14% (124)

change the correct power reference. In this case, the battery storage bank operates in the discharge mode in order to respond to the load demand.

11.4 Conclusions

In this chapter, a study on the control structure of autonomous hybrid energy generation system is performed. For possible implementation on a FPGA platform, the centralized PMS and local controllers of both renewable generators and battery bank are constructed using the XSG tool. The fixed-point format is so adopted to represent the system data and to perform different mathematical operations. The simulation results show satisfactory performance of the digitalized control scheme under different conditions of operation. Minimal quantization errors and high order of precision are attained with the control structure established using the XSG tool. These results demonstrate the correctness of the XSG design that can be tested now in real time using a FPGA card without risk of damage of the power system.

References

1. Chettibi, N., Mellit, A.: Intelligent control strategy for a grid connected PV/SOFC/BESS energy generation system. Energy **147**, 239–262 (2018)
2. Han, Y., Chen, W., Li, Q., Yang, H., Zare, F., Zheng, Y.: Two-level energy management strategy for PV-Fuel cell-battery-based DC microgrid. J Hydrogen Energy. **44**(35), 19395–19404 (2019)
3. Serir, C., Rekioua, D., Mezzai, N., Bacha, S.: Supervisor control and optimization of multi-sources pumping system with battery storage. J Hydrogen Energy **41**(45), 20974–20986 (2016)
4. Kalantar, M., Mousavi S. M.G.: Dynamic behavior of a stand-alone hybrid power generation system of wind turbine, microturbine, solar array and battery storage. Appl. Energy. **87**, 3051–3064 (2010)
5. Ouachani, I., Rabhi, A., Yahyaoui, I., Tidhaf, B., Fernando Tadeo, T.: Renewable energy management algorithm for a water pumping system. In: 8th Interrernational Conference Sustainability in Energy and Buildings, Energy Procedia, vol. 111, pp. 1030–1039 (2017)

6. Boukettaya, G., Krichen, L.: A dynamic power management strategy of a grid connected hybrid generation system using wind, photovoltaic and Flywheel Energy Storage System in residential applications. Energy **71**, 148–159 (2014)
7. Hu, J., Shan, Y., Xu, Y., Guerrero, J.M.: A coordinated control of hybrid ac/dc microgrids with PV-wind-battery under variable generation and load conditions. Electr. Power Energy Syst. **104**, 583–592 (2019)
8. Sharma, R.K., Mishra, S.: Dynamic Power Management and Control of PV PEM fuel Cell based Standalone AC/DC Microgrid Using Hybrid Energy Storage. IEEE Trans. Ind. Appl. **54**(1), 526–538 (2018)
9. Abadlia, I., Bahi, T., Bouzeria, H.: Energy management strategy based on fuzzy logic for compound RES/ESS used in stand-alone application. J. Hydrogen Energy **41**(38), 16705–16717 (2016)
10. Madaci, B., Chenni, R., Kurt, E., Hemsas, K.E.: Design and control of a stand-alone hybrid power system. J. Hydrogen Energy **41**(29), 12485–12496 (2016)
11. Pu, Y., Li, Q., Chen, W., Liu, H.: Hierarchical energy management control for islanding DC microgrid with electric-hydrogen hybrid storage system". J. Hydrogen Energy **44**(11), 5153–5161 (2018)
12. N.E. Zakzouk, M. A. Elsaharty, A. K. Abdelsalam, A.A. Helal,B. W. Williams, "Improved performance low-cost incremental conductance PV MPPT technique » IET Renew. Power Gener., Vol. 10, Iss. 4 (2016) pp. 561–574
13. Harrouz, A., Benatiallah, A., Moulayali, A., Harrouz, O.: Control of Machine PMSG Dedicated to the Conversion of Wind Power Off-Grid. In: 4th International Conference on Power Engineering, Energy and Electrical Drives Istanbul, Turkey, 13-17 May 2013
14. Meghni, B., M'Sirdi, N. K., Saadoun, A.: A novel maximum power tracking by VSAS approach for permanent magnet direct drive WECS. In: 7th Inter. Conf. Sustainability in Energy and Buildings, Energy Procedia vol. 83, pp. 79–90 (2015)
15. Haque, M. E., Muttaqi, K. M., Negnevitsky, M.: Control of a stand alone variable speed wind turbine with a permanent magnet synchronous generator. In: 2008 IEEE Power and Energy Society General Meeting-Conversion and Delivery Of Electrical Energy in 21st century

Chapter 12
Photovoltaic Model Based on Manufacture's Datasheet: Experimental Validation for Different Photovoltaic Technologies

R. Boukenoui, A. Mellit, and A. Massi Pavan

Abstract In this chapter, three different commercialized photovoltaic (PV) technologies—polycrystalline silicon (poly C–Si), copper indium gallium selenide (CIGS) and cadmium telluride (CdTe)—are investigated in terms of several aspects. A PV model based on manufacture's datasheet has been presented. Its originality consists in the using of a simple procedure which takes only the datasheet parameters into account to identify the series resistance (Rs) of solar cells. Moreover, the ideality factor (n) value is adapted to fit the solar cell technology. Both the identified Rs and n values have been used within the solar cell block provided by MATLAB Simscape toolbox to model different PV modules having different technologies, as well as to predict their characteristics (current–voltage (I–V) and Power–Voltage (P–V)). A test facility is employed to carry out the required tests for assessing the PV model. Obtained experimental results under different climate conditions are compared with simulated ones. The comparison is carried out by evaluating four statistical errors with a view of measuring the accuracy of the proposed model in predicting the I–V and P–V characteristics.

Keywords Photovoltaic technology · Crystalline silicon · Thin film · Simscape · Conversion efficiency · Photovoltaic modeling

R. Boukenoui (✉)
Faculty of Technology, Blida 1 University, 09000 Blida, Algeria
e-mail: rachidboukenoui@gmail.com

A. Mellit
Renewable Energy Laboratory, Jijel University, 18000 Jijel, Algeria

A. Massi Pavan
Department of Engineering and Architecture, University of Trieste, Via A. Valerio, 6/A, 34127 Trieste, Italy

© The Editor(s) (if applicable) and The Author(s), under exclusive license to Springer Nature Switzerland AG 2020
A. Mellit and M. Benghanem (eds.), *A Practical Guide for Advanced Methods in Solar Photovoltaic Systems*, Advanced Structured Materials 128,
https://doi.org/10.1007/978-3-030-43473-1_12

12.1 Introduction

Substantial rise in global energy demand together with the continuous depletion of conventional energy sources are motivating researches around the word seeking sustainable, green and more efficient energy sources. Green energy sources seem to be considered as an effective optimal pathway for future development efforts and efficient solution to cope with the environmental problems caused by conventional sources such as air pollution and global warming [1].

The primary energy source of photovoltaic (PV) systems is the sunlight, and its conversion to electric energy involves neither moving parts nor fuel burning. Various advantages such as noiseless, maintenance free, inexhaustible and environmentally friendly will stand out by using PV-based conversion system [2]. However, its main drawback, that is initial installation cost, is being addressed by extensive researches focusing on developing cheap PV technology with acceptable conversion efficiency. This leads to a continuous decrease in the average PV module [3]. According to a recent study conducted by Bloomberg New Energy Finance (BNEF), typical utility-scale PV systems were estimated at being 25% cheaper per MW in 2017 than they were in the year of 2015. This makes the number of installed PV plants and investments across the word rapidly increasing. In 2017, China has already invested $86.5 billion to improve its solar power generation. This was 58% higher than in 2016, with an estimated of 53 GW of PV capacity installed—up from 30 GW in 2016. Moreover, 2.1 GW of new PV capacity was installed in Germany [4], which corresponds to around 2% of total new PV capacity worldwide. It is worth mentioning that today, China still dominates with an estimated of 65 GW of PV power generation. With reference to IEA, the overall capacity of PV installations around the world is about 500 GW [5]. In 2019, around $9 billion has been invested by the global PV industry, with Chinese manufacturers claiming a share of around 85% regarding PV modules production capacity expansions. Such investments will further drive down cost, meanwhile, stimulating demand around the world in future.

Depending on the location with respect to the equator and the environmental conditions, the amount of solar energy harvesting per square meter can reach 1 kW. In some regions where solar irradiance rates are high while wide area of land is available (e.g., South Asia, Southeast Asia and Sub-Saharan Africa), if that amount of energy could be converted directly into electricity with 100% efficiency, it would be a real solution to the current world's energy consumption and the future energy demand [6]. Solar irradiance can be used directly through solar cells. Based on PV effect, the solar cells produce DC electricity when they are illuminated. Due to the obvious advantages of PV sources, various applications based on PV generation systems have been put into practice, starting from powering satellites to domestic water heating. PV plant can be used either as a standalone PV system (i.e., where grid utility is not available) or as grid-connected PV source. Currently, different PV technologies are available due to global efforts in research and development on the subject [7]. The most common PV technologies are crystalline silicon (C–Si) and thin films. However, each technology presents merits and demerits from different aspects including technical and economic.

Modeling of PV generator (which can be solar cell, module, string or array) is central for design, use and control. An accurate model allows (1) predicting the energy yield of PV generators having different configurations of PV modules (e.g., series-parallel, bridge-linked, total-cross-tied, etc.) [8], (2) evaluating the performance of maximum power point trackers [9], (3) performing loss analyses under static and dynamic environmental conditions (e.g., calculating both static and dynamic efficiencies of the whole PV system under different climatic conditions), (4) analyzing the integration of PV generations to AC and DC distribution systems [10]. Many techniques have been proposed in the literature to model PV generators. For instance, the techniques proposed in [11] are developed for determining the I–V and P–V characteristics. However, those characteristics are not immediately determined as additional steps are required for the computation of several parameters [12]. Sometimes the calculated parameters are sensitive to the climate conditions (e.g., the air humidity) as in [13]. The iterative method presented in [14] is sensitive to the initial solution which results in convergence failure.

Furthermore, even good initial guess cannot ensure convergence. Straightforward methods have the advantage to require less computation, and an example of these methods is developed in [15]. They are based on an analytical function that depends on the solar cell temperature and the irradiance level to compute current, voltage and power at the maximum power point (MPP). It can be noticed that such methods simplify the computation by avoiding the iterative process, but the accuracy is much lowered. Some models ignore the internal resistance to simplify the computation, and the obtained results show that accuracy is acceptable as in [16]. Other techniques have been developed to handle the fact that some PV parameters such as efficiency, fill factor and open-circuit voltage are climate change dependent. To this, polynomial regression [17] and artificial neural networks (ANN) have been investigated [18]. Both approaches are typically based on the power rating as input parameter. In addition, the advantage arises from the fact that they can be employed without the need for detailed knowledge about the system under study.

However, the disadvantage of using such approaches is that a large database of climatic and electrical data is needed to be collected in order to model the system. Genetic Algorithm (GA) [19], Particle Swarm Optimization (PSO) [20], Differential Evolution (DE) [21] and Artificial Neural Network (ANN) [22], etc., have been also employed to find the equivalent circuit parameters. A critical review of various modeling approaches including analytical models, classical optimization methods and soft computing techniques under different working conditions is presented in [23]. In the other hand, different softwares are used for the implementation and simulation of PV generators such as MATLAB, Maple, Pspice and SABER.

In light of the previous analysis, one can find that PV model should be at once efficient, accurate and easily to be applied by designers. Moreover, it is much better if the developed PV model uses only the data provided by the manufacture to model any configuration of PV parts (solar cells, bypass diodes, etc.). For this purpose, the present work describes a simple PV model based only on the manufacture's datasheet to model different PV modules having different technologies. It consists to use a simple procedure to identify the series resistance of solar cells, in addition to the ideality factor value which is adapted to fit the solar cell technology. The PV model uses MATLAB Simscape toolbox (blocks of solar cells, bypass diodes and physical networks) that offers the freedom to modify the configuration of solar cells and bypass diodes by playing on the physical connections between them.

The chapter is organized as follows, Sect. 12.2 discusses the recent progress and the challenges associated with C–Si and thin-film technologies. Section 12.3 presents the PV model for predicting the I–V and P–V characteristics of different PV module technologies. In Sect. 12.4, several experimental test results accompanied with several statistical error analyses, to validate experimentally the described PV model, are presented.

12.2 PV Module Technologies

Nowadays, solar cells with C–Si or thin-film technologies are very familiar from use viewpoint. They comprise a wide portion of installations worldwide. However, both technologies face challenges and present merits and demerits. These challenges include materials optimization, efficiency improvement and cost reduction. Therefore, in the following, the C–Si and thin-film technologies are compared in terms of various parameters. Since the release of C–Si and thin-film technologies to the market, research seeking for better efficiency has never been stopped. The evolution of efficiency improvement for each PV technology since 2009 can be found in [24].

12.2.1 Crystalline Silicon (C–Si) Technologies

The leader in PV industry and applications is C–Si technology. Silicon is a material with a gap energy of 1.1 eV, which be found in more than 90% of PV production [25]. PV modules with C–Si solar cells have been the focus of research since the 1950s, and currently, they cover around 85% of the PV market [26]. Depending on a long-term field testing for reliability and a high volume manufacturing facilities, C–Si PV industry is regarded as the mature one. The theoretical maximum efficiency for this technology is about 29% [27], with world record efficiency of 26.1 and 22.8% for mono C–Si and poly C–Si, respectively [28]. This technology is expected to survive for a multiple decades due to its significant merits in economic and technical aspects. According to Fig. 12.1, mono C–Si technology retains the highest efficiency in comparison with thin-film technologies (i.e., based on either CdTe or CIGS). Furthermore, the rise in efficiency for C–Si (poly C–Si and mono C–Si) and CIGS is low with respect to that of CdTe.

Poly C–Si solar cells are the next class of C–Si technologies. In fact, the major advantage of poly C–Si is the cost reduction compared to mono C–Si class. However,

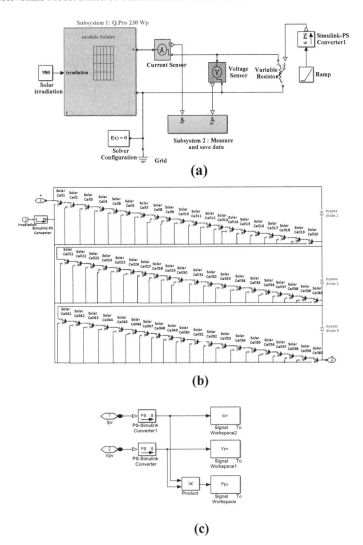

Fig. 12.1 Simscape implementation of: **a** 'Q.Pro 230Wp' PV module **b** PV system **c** integrated blocks in subsystem 2

this cost reduction resulted in a low conversion efficiency. Poly C–Si benefits from the maturity of the C–Si industry. Thereby, it is cheaper in the market as its fabrication process is faster and much easier than the other existing silicon technologies [29]. The average price of PV modules with poly C–Si technology is down about 14% per year (US$ 0.6/Watt achieved in 2014) [30]. In the last decade, poly C–Si and mono C–Si account for most of the market shares in PV manufacturing industry. Furthermore, poly C–Si PV modules are widely used in all kind of applications thanks to their trade-off between the low cost and the average power efficiency [31]. However, it is worth noting that poly C–Si and mono C–Si suffer from some limitations: (1) the relative high cost involved in the fabrication compared to thin films, (2) the weight and rigidity of modules, (3) the poor performance under high temperature and low insolation levels [32].

12.2.2 Thin-Film Technologies

Great interest has been devoted to the investigation of new alternative materials and device processing technologies in order to overcome some of the C–Si technology limitations. This has leaded to new technologies which expand rapidly, especially, thin-film technologies.

As reported in [31], thin-film technologies' production in 2015 has been increased by an estimated 25%. Nowadays, these technologies are held the most remainder portion of the PV market after C–Si.

The contributions of thin film lie in the no need for expensive semiconductor substrates and the reduced amount of active semiconducting material used in each solar cell. In manufacturing of thin-film solar cells, the used film thicknesses are typically around hundreds of nanometers to a few micrometers to absorb nearly all the incoming insolation. On the other hand, the thickness required in C–Si technology solar cells is about hundreds of micrometers; this resulted in the price advantage of thin-film solar cells. Moreover, thin films can be deposited onto a variety of inexpensive supporting substrates such as plastic, glass and foil.

Thin-film technologies are commercially available on the PV market either cadmium telluride (CdTe), copper indium gallium selenide (CIGS) or amorphous silicon. This latter is not sufficient to offset the advantages of the other thin-film technology; therefore, it is rapidly receding.

12.2.3 Cadmium Telluride CdTe

CdTe regarded as one of the first semiconductors that was used in thin-film technologies to improve the low efficiency experienced with amorphous silicon. Furthermore, the tremendous research targeting CdTe is to create a balance between performances and cost, making it essential in achieving high conversion efficiencies. The CdTe is a robust semiconducting material, and this is because it has a remarkable tolerance toward the high processing temperatures [25]. At present, PV modules with CdTe-based thin-film technology are one of the fastest-growing segments of commercial module production [33]. Therefore, CdTe-based thin-film technologies now firmly established as the basis for the market-leading thin-film PV module technology as well as the most cost effective to manufacture. As discussed in [34], the cost of CdTe PV modules has now been dropped below one US$-per-watt. As a result, the cost of power generation is rapidly approaching grid parity.

The glass substrate plate that layered the CdTe compound has a high optical absorption coefficient in order to enable light to reach the semiconductor layer more efficiently; as a result, only a low level of irradiance is needed to produce power [32]. PV modules based on CdTe technology are commercially available nowadays with conversion efficiency ranged between 7% and 12% on average. However, the tremendous improvements of this technology have increased its cell efficiency record to 22.1%.

From Fig. 12.1, the path toward high efficiency for CdTe has been quicker than the other considered technologies, which results in latest achieving efficiency almost equal to that of poly C–Si one.

It is important to note that CdTe-based thin-film solar cells have a simple structure of the two-component absorber layer (i.e., contain only cadmium and tellurium). However, tellurium has a limited supply, which makes CdTe-based thin-film PV modules not as a real solution for large-scale deployment. Likewise, the cadmium is a toxic material which is dangerous to both industrial operators and outdoor applications. Therefore, it must be treated carefully in its handling and disposal. Consequently, further improvements in conversion efficiency as well as in material and interface properties of CdTe solar cells are needed.

12.2.4   Copper Indium Gallium Selenide, CIGS

CIGS technology is regarded among the most popular thin-film technologies. It offers inherent advantages of thin film for cost reduction and high absorption coefficients (i.e., with a band gap of 1.68 eV). Copper, indium, gallium and selenium [Cu (In, Ga) Se$_2$] are the main materials that compose this semiconductor type.

A key stage in the fabrication of CIGS solar cells is to choose the appropriate substrate in order to create a compromise between the desired performances and the processing conditions. Thin CIGS films can be deposited onto a variety of flexible (e.g., mostly metals, ceramics, special polymers, etc.) and rigid (e.g., soda lime glass, etc.) substrate materials. However, the used substrate material has to be chosen considering its availability in large quantities, the compatibility with the subsequent deposition steps (i.e., withstanding to different chemical and physical properties as well as to the high temperature level during the manufacturing process), cost effective, robustness and lightweight. The considered technique that still under way for the development to produce large commercialization of CIGS PV modules on flexible substrates, is the roll-to-roll (R2R) production technique. The moderate efficiency associated with CIGS-based thin-film PV modules ranges from 10 to 15%, with record conversion efficiency of 23.4% [28].

Applying a flexible substrate to CIGS thin film allows obtaining a large-scale flexible PV plant and opens new fields of applications (e.g., coated on the existing surfaces or integrated with building components such as roof tiles).

By comparing C–Si and thin films, C–Si technologies are the most suitable for the large-scale deployment considering their abundance and the ratio efficiency to price, specifically, mono C–Si PV technology which is more efficient but relatively more expensive than poly C–Si one. However, in hot places, CIGS or CdTe is more recommended.

12.3 Photovoltaic Modeling

An improved PV model based on MATLAB Simscape toolbox for obtaining the I–V and P–V characteristics is developed. The modeling tools consist of physical component blocks (i.e., solar cell blocks, bypass diode blocks, etc.) and physical networks. In Simscape environment, solar cells and bypass diodes can be seamlessly used to model PV modules (i.e., built by interconnecting individual solar cells), PV strings (i.e., built by connecting PV modules in series to get a sufficient voltage) or arrays (i.e., built by connecting PV strings in parallel to get a sufficient current) [35]. Simscape physical networks are employed to connect together any configuration of PV parts as well as to transmit power between them.

The solar cell block provided by Simscape is comprised of a parallel combination of current source, a single diode and parallel resistor (R_p); the latter is connected in series with a resistance R_s. The solar cell current I is given by

$$I = I_{ph} - I_0 \left(\exp\left(\frac{q(V + R_s I)}{KnT} \right) - 1 \right) - \frac{(V + R_s I)}{R_p} \tag{12.1}$$

where V and I represent the solar cell output voltage and current, respectively. R_s is the series resistance of a single solar cell, which has usually a small value representing the contact resistance. However, the parallel resistance R_p possesses a very large resistance that models the leakage current of the P-N junction. q [C] is the charge of an electron (1.602×10^{-19}). I_{ph} is the light current, I_0 is the saturation current, n is the diode ideality factor, K [J/K] the Boltzmann's constant (1.38×10^{-23}) and T is the cell temperature. So, to characterize a solar cell block, **the block-parameters of solar cell** require both climatic and electrical parameters are required. The climatic parameters which are the temperature [°C] and the irradiance [W/m^2] can be obtained by measuring them with the appropriate sensors. The electrical parameters are the open-circuit voltage (V_{OC}), the short-circuit current (I_{SC}), the ideality factor n and the series resistance R_s. The V_{OC} and I_{SC} values are obtained from PV module's datasheet. However, n and R_s are not provided.

The previous works to obtain the series resistance by using trial and error method are presented in [36, 37]. However, to model a PV plant with different parameters of PV modules, it is a difficult task to find an accurate R_s for each module. In [38], R_s is determined by evaluating the derivative of the model equation at V_{OC} point, which leads to significant computation efforts. An iterative algorithm is used in [39] that is based on the minimization of (dP /dV$_{MPP}$ = 0) to find an optimal value of the adjustment factor (α). Once this latter is reached, the R_s can be calculated. Another approach based on optimization is presented in [40] where the determination of R_s

Table 12.1 Ideality factor for different PV technologies

Technology	Ideality factor
Poly C–Si	1.3
CdTe	1.5
CIGS	1.5

value is carried out by evaluating an expression based on unknown factor and the parameters reported in the datasheet. However, this approach is very sensitive to initial solution as it uses the Newton–Raphson method to determine an unknown factor within the expression of R_s. In the same context, the authors in [41] have employed a binomial search routine to seek the optimal value of R_s within an estimated interval. However, the use of this optimization method made the determination of R_s a complex task.

A complicated function of series resistance is solved iteratively in [42] by using a typical series resistance of the PV module as a starting point. However, the model needs additional information besides the standard data provided by the manufacturer's datasheet. Considering the aforementioned drawbacks, designers of PV systems often find difficulties in applying such models. On the other hand, in many modeling techniques, the ideality factor is not adapted to fit the modeled PV technology. Therefore, it is taken as a random value from the range $1 < n < 2$.

In the following, a contribution for rendering the Simscape-based PV model based only on the manufacture's datasheet is made. Therefore, for each technology of PV modules, a specific ideality factor value is used. Poly C–Si, CdTe and CIGS ideality factors are given in Table 12.1. Moreover, an expression uses the available information in the manufacturer's datasheet to forwardly calculate the value of series resistance R_s which has been adopted.

The R_s calculation is based on the computation of the fill factor which is determined as follows:

$$V\_{\text{MPP}}[p.u] = \frac{V_{\text{MPP}}}{V_{\text{oc}}} \tag{12.2}$$

$$I\_{\text{MPP}}[p.u] = \frac{I_{\text{MPP}}}{I_{\text{sc}}} \tag{12.3}$$

$$\text{FF}[] = V\_{\text{MPP}}[p.u].I\_{\text{MPP}}[p.u] \tag{12.4}$$

where V_{OC} 9 [V] and I_{SC} [A] are the open-circuit voltage and the short-circuit current, respectively. FF [] is the PV module's fill factor at STC.

The solar cell's series resistance R_s [Ω] [43] is calculated as:

$$R_S = \frac{V_{\text{OC}}}{I_{\text{SC}}}.R_{\text{SN}} \tag{12.5}$$

R_{SN} [] is the normalized solar cell's series resistance that is determined as shown below [44]

$$R_{SN} = 1 - \frac{FF}{FF_{N,25}} \tag{12.6}$$

where $FF_{N,25}$[] is the normalized fill factor when the solar cell temperature is 25 °C and given as:

$$FF_{N,25} = \frac{v_{OCN,25} - \ln(v_{OCN,25} - 0.72)}{v_{OCN,25} + 1} \tag{12.7}$$

where $V_{OCN,25}$ [] is the normalized open-circuit voltage at temperature of 25 °C which is calculated as [43]:

$$v_{OCN,25} = \frac{V_{OC}}{Vt_{,25}.N_S} \tag{12.8}$$

where $V_{t,25}$[V] is the junction thermal voltage of the solar cell at 25 °C, and N_s [] is the number cells of the PV module.

Finally, the junction thermal voltage of the solar cell at 25 °C is given by;

$$Vt_{,25} = \frac{k.(25 + 273)}{q} \tag{12.9}$$

where k [J.K^{-1}] and q [C] are the Boltzmann's constant and the electron charge, respectively.

Figure 12.1a shows the Simscape implementation of the PV module 'Q.Pro 230Wp' connected to a system for sweeping acquiring the I–V and P–V characteristics; its main partis:

- **Subsystem 1**: contains the implemented Q.Pro 230Wp PV module illustrated in Fig. 12.1b; it is composed of 60 solar cells connected together by physical networks. Each solar cell has been characterized according to the electrical characteristics reported in Table 12.2 as well as the ideality factor values reported in Table 12.1. Equations (2–9) have been used to compute the solar cell's series resistance.
- **A variable resistor and ramp block**: used to sweep the I–V and P–V characteristics.
- **Current and voltage sensors**: employed to measure I and V, respectively. The measured I and V are acquired and saved by Subsystem 2.
- **Subsystem 2**: used to acquire and save the measured data. Its main blocks are shown in Fig. 12.2c.
- **Solver Configuration block**: used to determine the unknown variables for the entire modeled system; it can be connected anywhere on the physical network

Table 12.2 Electrical characteristics of the used PV modules

Designation	Q.Pro 230Wp	First Solar FS-272	Q.Smart 95Wp
Maximum power (P_{MPP}) [W]	230	72.5	95
Voltage at P_{max}(V_{MPP}) [V]	29.24	66.6	62.1
Current at P_{max} (I_{MPP}) [A]	7.95	1.09	1.53
Short-circuit current (I_{SC}) [A]	8.59	1.23	1.68
Open-circuit voltage (V_{OC}) [V]	36.95	88.7	78
Cells number N_S (cells)	60	116	116
Bypass diodes number	3	None	1
Technology	Poly C–Si	CdTe	CIGS

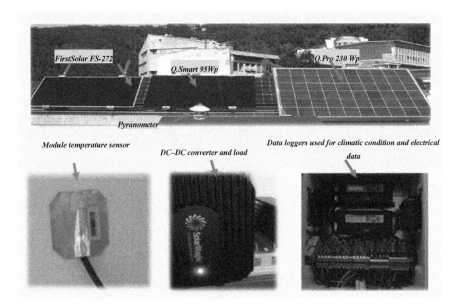

Fig. 12.2 Test facility to measure the electrical characteristics of the used PV modules at Trieste University, Trieste, Italy

circuit by creating a branching point. It is imperative to fine-tune the solver parameters before running the simulation in order to avoid warning or error message. The solver parameters are chosen as follows: the step type is *fixed step*, the solver type is **Runge–Kutta** and fixed step size is 0.001.

12.4 Results and Discussions

12.4.1 Experimental Data Analysis

A number of measurements have been carried out at the test facility of Trieste University, Italy. To this, the three PV modules, namely Q.Pro 230Wp (Poly C–Si technology), Q.Smart 95Wp (CIGS technology) and First Solar FS-272 (CdTe technology), have been considered. The electrical specifications of these modules are reported in Table 12.2. Different experiments are conducted at different working conditions for each of the above modules. The I–V characteristic of each PV module is recorded by sweeping the output voltage from zero to V_{OC}.

The test facility is shown in Fig. 12.2, where three PV modules are made available. It comprises also two data loggers (type E-Log, MW8024-02/10 produced by LSI Lastem S.r.l) for climatic and electrical data; they are connected to a computer for the data to be collected and saved. DC–DC converter (type Solar Magic produced by National Semiconductor Ltd) connected to a resistive load. One ISO9060 first class thermopile global radiometer type C100R DPA153 (produced by LSI Lastem S.r.l) installed on the same frame carrying PV modules(the daily uncertainty for this device is less than 5%, the sensitivity is $30 \div 45 \ \mu V/(W/m^2)$ and the flat spectral response range is (305–2800 nm)). Three module temperature contact probes (type DLE 124 by LSI Lastem S.r.l) have an accuracy of ± 0.15 °C. Three shunts type SHP300A60-Compact (produced by Hobut Ltd) calibrated with an accuracy better than 0.01%. Fast measurements of the I–V characteristics are carried out by varying the duty cycle of the DC–DC converter from the minimum to its maximum value. At this time, both the climatic (irradiance and temperature) and electrical data (PV current and voltage) are simultaneously acquired and stored in the computer.

12.4.2 Experimental Validation of Simscape-Based PV Model

The validation of the Simscape-based PV model is carried out to show its effectiveness in predicting the I–V and P–V characteristics for different PV technologies (i.e., poly C–Si, CdTe and CIGS). Therefore, First Solar FS-272, Q.Pro 230Wp and Q.Smart 95Wp models are implemented according to the electrical characteristics reported in Table 12.2 as well as the ideality factor values reported in Table 12.1. Equations (2–9) have been used to compute the solar cell's series resistance for each PV module. The obtained values (Rs_First Solar FS-272 = 0.1402 Ω,Rs_Q.Pro 230Wp = 0.0094 Ω,Rs_ Q.Smart 95Wp = 0.0565 Ω) are used to parameterize solar cells. Simulation is done using the recorded climate conditions. The obtained I–V and P–V characteristics of Simscape-based PV model have been compared to those measured for the same weather conditions.

Figures 12.3, 12.4, 12.5, 12.6, 12.7 and 12.8 show the simulated versus the measured data of the I–V and P–V characteristics for each PV technology. It can be

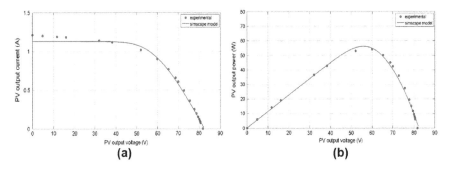

Fig. 12.3 Experimental data (dotted line), Simscape model (solid line) for PV module FS-272 operating at $G = 920$ W/m2 and $T = 56$ °C **a** I–V characteristic **b** P–V characteristic

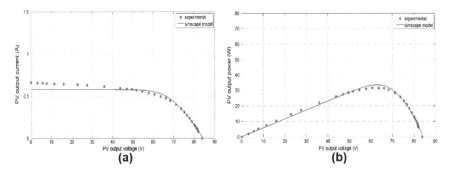

Fig. 12.4 Experimental data (dotted line), Simscape model (solid line) for PV module FS-272 operating at $G = 470$ W/m2 and $T = 30$ °C **a** I–V characteristic **b** P–V characteristic

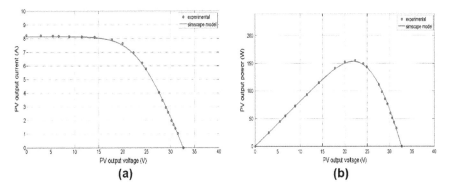

Fig. 12.5 Experimental data (dotted line), Simscape model (solid line) for PV module Q.pro operating at $G = 980$ W/m2 and $T = 55.5$ °C **a** I–V characteristic **b** P–V characteristic

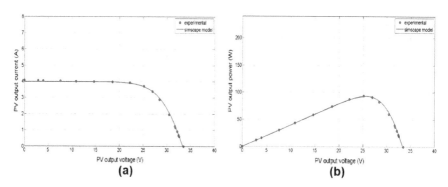

Fig. 12.6 Experimental data (dotted line), Simscape model (solid line) for PV module Q.pro operating at $G = 480$ W/m2 and $T = 36.5$ °C **a** I–V characteristic **b** P–V characteristic

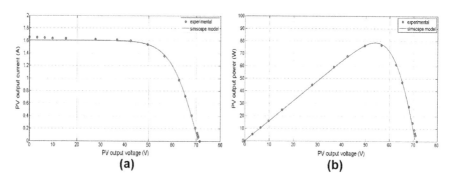

Fig. 12.7 Experimental data (dotted line), Simscape model (solid line) for PV module Q.smart operating at $G = 995$ W/m2 and $T = 56$ °C **a** I–V characteristic **b** P–V characteristic

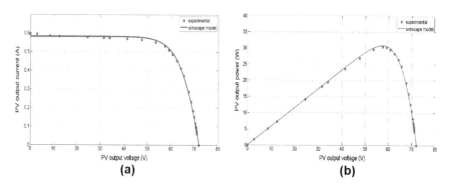

Fig. 12.8 Experimental data (dotted line), Simscape model (solid line) for PV module Q.smart operating at $G = 347$ W/m2 and $T = 30$ °C **a** I–V characteristic **b** P–V characteristic

noticed that there is a good agreement between the simulated and the measured electrical characteristics.

In order to measure the accuracy of the developed Simscape-based-PV model in predicting the P–V and I–V characteristics of each technology under the considered climatic conditions, several statistical error tests are used to characterize the degree of matching between the simulated and the measured characteristics. These statistical error tests are given by the following expressions:

The root mean square error (RMSE)

$$\text{RMSE}[W] = \sqrt{\frac{1}{N} \sum_{k=1}^{N} \left(P_{\text{pv\_mea}}(k) - P_{\text{pv\_sim}}(k) \right)^2} \qquad (12.10)$$

The correlation coefficient (R^2)

$$R^2[\%] = \left[1 - \frac{\sum_{k=1}^{N} \left(P_{\text{pv\_mea}}(k) - P_{\text{pv\_sim}}(k) \right)^2}{\sum_{k=1}^{N} \left(P_{\text{pv\_mea}}(k) \right)^2} \right] \times 100 \qquad (12.11)$$

The mean percentage error (MPE)

$$\text{MPE}[\%] = \frac{1}{N} \sum_{k=1}^{N} \left(\frac{P_{\text{pv\_sim}}(k) - P_{\text{pv\_mea}}(k)}{P_{\text{pv\_mea}}(k)} \right) \times 100 \qquad (12.12)$$

The deviation from the measured values (Dev)

$$\text{Dev}[] = \frac{\sum_{k=1}^{N} P_{\text{pv\_sim}}(k) - \sum_{k=1}^{N} P_{\text{pv\_mea}}(k)}{\sum_{k=1}^{N} P_{\text{pv\_mea}}(k)} \qquad (12.13)$$

where

$P_{\text{pv\_mea}}(k)$ [W] is the kth measured value of the PV output power, $P_{\text{pv\_sim}}(k)$ [W] is the kth simulated value of the PV output power and N [] is the number of simulated or measured points.

Table 12.3 shows, for each PV module and for given climate conditions, the computed statistical errors. One can notice that the RMSE values are relatively low, ranging between 0.311 and 1.246 W. Furthermore, the deviations from the measured values are the range that is [−0.0125, 0.0118]. The correlation coefficient values are greater than 99.8% regardless the technology of the module and the weather conditions. Moreover, the absolute values of the MPE are always less or equal to 0.057%. Upon the analysis of these results, it can be concluded that the proposed

Table 12.3 Comparisons between simulated and measured P-V curves of each used PV technologies

PV module	Technology	G (W/m^2)	T (°C)	RMSE (W)	R^2 (%)	MPE (%)	Dev []
First Solar FS-272	CdTe	920	56	1.2462	99.82	−0.057	−0.0125
		470	30	0.7937	99.85	0.0357	0.0118
Q.Pro 230Wp	Poly C–Si	980	55.5	0.5976	99.99	0.0031	0.00064
		480	36.5	0.5762	99.99	0.0236	0.0045
Q.Smart 95Wp	CIGS	995	56	0.5767	99.98	−0.0123	−0.0022
		347	30	0.3118	99.97	0.0185	0.0050

Simscape-based PV model offers a good performance in predicting the electrical characteristics (i.e., current, voltage, power) of different PV technologies, particularly, a PV module having poly C–Si technology. The latter has the highest correlation values (99.99%) and the lowest |MPE| and |Dev, 0.0031 and 0.00064, respectively. Likewise, the obtained static errors of CIGS PV module lead to say that Simscape-based PV model can offer satisfactory prediction for this technology. However, the predicted electrical characteristics of CdTe PV module present the poorest performances compared to the remaining. This has the lowest correlation (99.82%) and the highest |MPE| and |Dev| (0.057 and 0.0125, respectively).

12.5   Conclusion

In this chapter, a survey of the recent progress on the most commercialized PV technologies (i.e., C–Si and thin film) has been presented. Mono C–Si technology is the leader in both PV applications, industrial and building integrated. Furthermore, it is considered as the most suitable for the large-scale deployment. On the other hand, thin-film technology suffers from the low conversion efficiency. However, it is expected to be increased in the future. Turning now to the second part of this chapter where a simple PV model based on Simscape for predicting the electrical characteristics of different PV technologies has been implemented. Then, it has been experimentally tested under different working conditions. Assessment criteria based on four statistical errors have been calculated to precisely evaluate the Simscape-based PV model. It has been shown that the presented Simscape-based PV model offers a satisfactory accuracy of prediction. This model is more appropriate for users due to its simplicity, and it can be implemented easily.

Acknowledgements The authors would like to thanks Dr. V. Lughi, from Trieste University, Italy, for the useful database.

References

1. Sarhaddi, F., Farshchi Tabrizi, F., Aghaei Zoori, H., et al.: Comparative study of two weir type cascade solar stills with and without PCM storage using energy and exergy analysis. Energy Convers. Manag. **133**, 97–109 (2017)
2. Bhattarai, S., Kafle, G.K., Euh, S.-H., et al.: Comparative study of photovoltaic and thermal solar systems with different storage capacities: performance evaluation and economic analysis. Energy **61**, 272–282 (2013)
3. Caballero, F., Sauma, E., Yanine, F.: Business optimal design of a grid-connected hybrid PV (photovoltaic)-wind energy system without energy storage for an Easter Island's block. Energy **61**, 248–261 (2013)
4. Bruno, B.: Fraunhofer Institute for Solar Energy Systems ISE, Freiburg, Germany. https://www.energy-charts.de
5. (IEA) IEA.: Renewables (2019)
6. Ma, T., Yang, H., Zhang, Y., et al.: Using phase change materials in photovoltaic systems for thermal regulation and electrical efficiency improvement: a review and outlook. Renew. Sustain. Energy Rev. **43**, 1273–1284 (2015)
7. Polo, J., Alonso-Abella, M., Ruiz-Arias, J.A., et al.: Worldwide analysis of spectral factors for seven photovoltaic technologies. Sol. Energy **142**, 194–203 (2017)
8. Dadje, A., Djongyang, N., Kana, J.D., et al.: Maximum power point tracking methods for photovoltaic systems operating under partially shaded or rapidly variable insolation conditions: a review paper. Int. J. Sustain. Eng. **9**(4), 224–239 (2016)
9. Kheldoun, A., Bradai, R., Boukenoui, R., et al.: A new golden section method-based maximum power point tracking algorithm for photovoltaic systems. Energy Convers. Manag. **111**, 125–136 (2016)
10. Gab-Su, S., Jong-Won, S., Bo-Hyung, C., et al.: Digitally controlled current sensorless photovoltaic micro-converter for DC distribution. IEEE Trans. Ind. Inform. **10**(1), 117–126 (2014)
11. Chatterjee, A., Keyhani, A., Kapoor, D.: Identification of photovoltaic source models. IEEE Trans. Energy Convers. **26**(3), 883–889 (2011)
12. Massi Pavan, A., Mellit, A., Lughi, V.: Explicit empirical model for general photovoltaic devices: experimental validation at maximum power point. Sol. Energy **101**, 105–116 (2014)
13. Brus, V.: On quantum efficiency of nonideal solar cells. Sol. Energy **86**(2), 786–791 (2012)
14. Lun, S-x, Du, C-j, Yang, G-h, et al.: An explicit approximate I-V characteristic model of a solar cell based on padé approximants. Sol. Energy **92**, 147–159 (2013)
15. Saloux, E., Teyssedou, A., Sorin, M.: Explicit model of photovoltaic panels to determine voltages and currents at the maximum power point. Sol. Energy **85**(5), 713–722 (2011)
16. Babu, B.C., Gurjar, S.: A novel simplified two-diode model of photovoltaic (PV) module. IEEE J. Photovoltaics **4**(4), 1156–1161 (2014)
17. Huld, T., Friesen, G., Skoczek, A., et al.: A power-rating model for crystalline silicon PV modules. Sol. Energy Mater. Sol. Cells **95**(12), 3359–3369 (2011)
18. Mellit, A., Sağlam, S., Kalogirou, S.: Artificial neural network-based model for estimating the produced power of a photovoltaic module. Renew. Energy **60**, 71–78 (2013)
19. Dizqah, A.M., Maheri, A., Busawon, K.: An accurate method for the PV model identification based on a genetic algorithm and the interior-point method. Renew. Energy **72**, 212–222 (2014)
20. Khanna, V., Das, B., Bisht, D., et al.: A three diode model for industrial solar cells and estimation of solar cell parameters using PSO algorithm. Renew. Energy **78**, 105–113 (2015)
21. Chin, V.J., Salam, Z., Ishaque, K.: An accurate modelling of the two-diode model of PV module using a hybrid solution based on differential evolution. Energy Convers. Manag. **124**, 42–50 (2016)
22. Mekki, H., Mellit, A., Salhi, H.: Artificial neural network-based modelling and fault detection of partial shaded photovoltaic modules. Simul. Model. Pract. Theory **67**, 1–13 (2016)
23. Jena, D., Ramana, V.V.: Modeling of photovoltaic system for uniform and non-uniform irradiance: a critical review. Renew. Sustain. Energy Rev. **52**, 400–417 (2015)

24. Green, M.A., Dunlop, E.D., Levi, D.H., et al.: Solar cell efficiency tables (version 54). Prog. Photovoltaics Res. Appl. **27**(7), 565–575 (2019)
25. Kumar, S.G., Rao, K.K.: Physics and chemistry of CdTe/CdS thin film heterojunction photovoltaic devices: fundamental and critical aspects. Energy Environ. Sci. **7**(1), 45–102 (2014)
26. Wolden, C.A., Kurtin, J., Baxter, J.B., et al.: Photovoltaic manufacturing: present status, future prospects, and research needs. J. Vacuum Sci. Technol. A Vacuum Surf. Films **29**(3), 030801 (2011)
27. Shockley, W., Queisser, H.J.: Detailed balance limit of efficiency of p-n junction solar cells. J. Appl. Phys. **32**(3), 510–519 (1961)
28. Best research cell efficiencies.: The National Renewable Energy Laboratory (NREL) (2019)
29. Jiang, Y., Shen, H., Pu, T., et al.: High efficiency multi-crystalline silicon solar cell with inverted pyramid nanostructure. Sol. Energy **142**, 91–96 (2017)
30. Murdock, H. E., Gibb, D., André, T., et al.: Renewables 2019 Global Status Report (2019)
31. Raturi, A. K.: Renewables 2016 global status report (2016)
32. Reinhard, P., Chirila, A., Blosch, P., et al.: Review of progress toward 20% efficiency flexible CIGS solar cells and manufacturing issues of solar modules. In: IEEE 38th Photovoltaic Specialists Conference (PVSC), vol. 2, pp. 1–9 (2012)
33. Burst, J.M., Duenow, J.N., Albin, D.S., et al.: CdTe solar cells with open-circuit voltage breaking the 1 V barrier. Nature Energy **1**(3), 16015 (2016)
34. Major, J., Treharne, R., Phillips, L., et al.: A low-cost non-toxic post-growth activation step for CdTe solar cells. Nature **511**(7509), 334 (2014)
35. Boukenoui, R., Ghanes, M., Barbot, J.-P., et al.: Experimental assessment of maximum power point tracking methods for photovoltaic systems. Energy **132**, 324–340 (2017)
36. Orioli, A., Di Gangi, A.: A procedure to calculate the five-parameter model of crystalline silicon photovoltaic modules on the basis of the tabular performance data. Appl. Energy **102**, 1160–1177 (2013)
37. Luque, A., Hegedus, S.: Handbook of photovoltaic science and engineering. Wiley, England (1996). 2003
38. Walker, G.: Evaluating MPPT converter topologies using a MATLAB PV model. J. Electr. Electron. Eng. Australia **21**(1), 49 (2001)
39. Xiao, W., Dunford, W. G., Capel, A. A.: Novel modeling method for photovoltaic cells. In: 2004 IEEE 35th Annual Power Electronics Specialists Conference (IEEE Cat. No. 04CH37551), IEEE, pp. 1950–1956 (2004)
40. Ulapane, N. N., Dhanapala, C. H., Wickramasinghe, S. M., et al.: Extraction of parameters for simulating photovoltaic panels. In: 2011 6th IEEE International Conference on Industrial and Information Systems (ICIIS), IEEE, pp. 539–544 (2011)
41. Chenni, R., Makhlouf, M., Kerbache, T., et al.: A detailed modeling method for photovoltaic cells. Energy **32**(9), 1724–1730 (2007)
42. Mäki, A., Valkealahti, S.: Power losses in long string and parallel-connected short strings of series-connected silicon-based photovoltaic modules due to partial shading conditions. IEEE Trans. Energy Convers. **27**(1), 173–183 (2011)
43. Massi Pavan, A., Mellit, A., De Pieri, D., et al.: A study on the mismatch effect due to the use of different photovoltaic modules classes in large-scale solar parks. Prog. Photovoltaics Res. Appl. **22**(3), 332–345 (2014)
44. Markat, T., Castañer, L.: Photovoltaics: fundamentals and applications. Preface Guide to Usage of the Handbook by Professional Group (2006)

Chapter 13
Recent Applications of Artificial Intelligence in Fault Diagnosis of Photovoltaic Systems

A. Mellit

Abstract This chapter presents a brief survey on the recent applications of artificial intelligence (AI) techniques in fault diagnosis of photovoltaic (PV) systems. AI-based methods are mainly used to identify and classify the type of faults that can happen in PV systems, particularly in DC side (PV array). The methods will be presented and discussed in terms of complexity implementation, possible faults detection, identification and localization capability. Faults localization in large-scale PV plants remains challenging issue, and to date, no AI-based method was applied and verified experimentally, except few methods recently developed based on aerial images (Infrared Thermography) inspection. It is believed that this chapter can help researchers in academic institutions to get an idea regarding the actual application of AI techniques in this topic.

Keywords Photovoltaic plants · Artificial intelligence techniques · Machine learning · Deep learning · Fault diagnosis

13.1 Introduction

As reported in [1], the installed PV capacity around the world at the end of 2018 was about 500 GW. The same source [1] indicated that all of the PV systems installed throughout the world are currently able to cover about 3% of global electricity demand. PV plants are subject to a number of different types of faults and failures (for example in the PV modules, in the wiring, inverter, and protection equipment), their yield can drop significantly, especially in desert regions (e.g., Sahara) where the operating conditions are extreme [2]. Thus, faults in PV systems may cause a huge amount of energy loss (Fig. 13.1a) as well as risk of fires (Fig. 13.1b), particularly

A. Mellit (✉)
Renewable Energy Laboratory, Faculty of Sciences and Technology, Department of Electronics, Jijel University, Jijel 18000, Algeria
e-mail: adelmellit2013@gmail.com

© The Editor(s) (if applicable) and The Author(s), under exclusive license to Springer Nature Switzerland AG 2020
A. Mellit and M. Benghanem (eds.), *A Practical Guide for Advanced Methods in Solar Photovoltaic Systems*, Advanced Structured Materials 128,
https://doi.org/10.1007/978-3-030-43473-1_13

(b)

(a)

Fig. 13.1 a Sandstorm effect on the PV arrays. **b** Fire hazards in a PV plant

in hot climate. In general, faults in PV arrays are mainly caused by dust, sand accumulation, mismatch, crack and ageing of PV modules. In order to keep the operation of a PV plant reliable, efficient and safety, a monitoring system (MS) including an automatic diagnosis method is indispensable [3].

Over the recent years many fault diagnosis techniques have been developed in the literature [4]. These techniques should be able to detect, identify and locate the position of the fault. For a small-size PV system, up to 10 kW (domestic application), it is relatively easy to diagnose the system using manual or semi-automatic methods. However, the localization of faults in large-size PV plants (utility-scale applications) is quite challenging issue.

The application of artificial intelligence (AI) techniques in PVs was shown the ability of such techniques to solve some problems such as: sizing and optimization (e.g., genetic algorithm (GA)), output power forecasting (e.g., neural networks (NNs)), control (e.g., ANNs and Fuzzy Logic (FL)) and maximum power tracking (FL) [5]. Nowadays, AI-based methods for fault detection, identification and localization in PV systems have attracted many researchers, due to the promising results of AI in this field. The main objective of this chapter is to present a short review on the recent application of AI techniques, including machine learning (ML) and recently deep learning (DL) in fault detection, identification and fault classification of PV systems.

The chapter is organized as follows: The next section presents a general introduction to machine learning and deep learning; Sect. 13.2 reports PV systems, including monitoring systems (MSs), different type of faults and available fault diagnosis methods. Application of AI techniques in fault detection, identification and classification is provided in Sect. 13.3.

13.2   Artificial Intelligence Techniques

There are many definition of AI, the most used definition is *'imitating intelligent human behaviour'*, which is already a much stronger definition, and it can classify [6]: (1) systems that think like humans, (2) systems that act like humans, (3) systems that think rationally and (4) systems that act rationally. Several intelligent computing technologies are becoming useful as alternate approaches to conventional techniques or as components of integrated systems. There are different branch of AI techniques and the main branch of AI used in PV applications are: Evolutionary algorithms (including different meta-heuristic methods such as Genetic Algorithm (GA), Particle Swarm Optimization and Ant Colony Optimization), machine learning (including support vector machine (SVM), k-nearest neighbour (k-NN), linear regression (LR), decision trees (DT), and naïve Bayes (NB)), NNs (including MLP, RNN and RBFN), FL, hybrid systems that combine two or more branch of AI (e.g., ANFIS, ANN-GA, etc.) and Deep learning (DL). Figure 13.2 shows the link between AI, ML and DL. As reported in [7], major progress in AI will come through systems that combine representation learning with complex reasoning.

13.2.1   Machine Learning

ML is the domain of study that gives computers the ability to learn without being explicity [7]. As shown in Fig. 13.2 ML is a subset of AI that can provide systems the

Fig. 13.2 Link between artificial intelligence, machine learning and deep learning

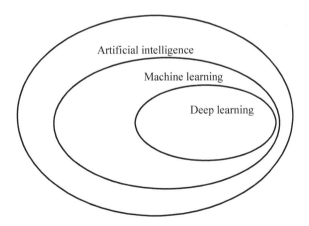

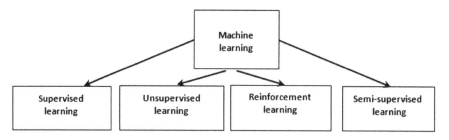

Fig. 13.3 Types of machine learning

ability to automatically learn and improve from experience without being explicitly programmed. It focuses mainly on the development of computer programs that can access data and use it to learn for themselves [7].

There are various ML algorithms such as linear regression, logistic regression, decision trees, naive Bayes, random forest, support vector machine, K-nearest neighbour, extreme learning machine, K-means clustering, Bayesian modelling and Q-learning. In point of view leaning, ML algorithms can be classified into four major algorithms (see Fig.13.3) [8]: (1) supervised learning, (2) unsupervised learning, (3) reinforcement learning and (4) semi-supervised learning. The first type of the algorithm tries to model relationships and dependencies between the target prediction output and the input features.

In the second type (unsupervised learning), no training labels for the training samples, the algorithm try to use techniques on the input data to mine for rules, detect patterns and group the data. It is mostly used to project high-dimensional data into low-dimension for visualization purposes.

Reinforcement learning has an agent that learns how to behave in an environment by taking actions and quantifying the results. If the agent makes a correct response, it gets a reward point, which boosts up the agent's confidence to take more such actions. The last type of learning (semi-supervised) is a combination between supervised and unsupervised, it consists of a small amount of labelled data and a large amount of unlabeled data, so labelled data are used for supervised, while unlabeled data for unsupervised learning during the training phase.

13.2.2 Deep Learning

It can be defined as seeking to provide knowledge to computers through data, observations and interacting with the world [9]. That acquired knowledge allows computers to correctly generalize to new settings. As shown in Fig. 13.2, DL is part of a larger family of ML methods based on learning data representations. DL is a relatively new advancement in ANN programming and represents a way to train DNNs, as traditional ANN method suffered from problems such as over-fitting, diminishing gradients [8]. As defined in [10], an ANN is a massively parallel-distributed processor made up of simple processing units that has a natural propensity for storing experiential knowledge and making it available for use.

In the last few years, DL has led to very good performance on a variety of problems, such as visual recognition speech recognition, natural language processing, pattern recognition, automatic translations, self-driving cars, medical diagnosis and financial prediction, automatic trading, but the application of DL in PV systems is very limited. The main DL algorithms used in PV applications are: (1) deep convolutional neural network (DCNN) and long short-term memory (LSTM).

- *Deep Convolutional neural network* is a specialized kind of ANN for processing data that has a known grid-like topology. DCNNs are simply ANNs that use convolution in place of general matrix multiplication in at least one of their layers [10]. Figure 13.4 shows a general architecture of DCNN.

- *Long short-term memory* is a kind of RNNs, the main drawbacks of RNNs practically fail to handle long-term dependencies. The LSTM [6] is capable of handling

Fig. 13.4 Architecture of a Deep learning neural network

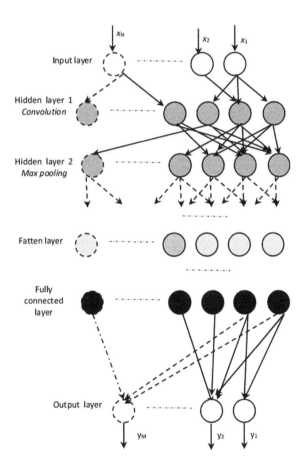

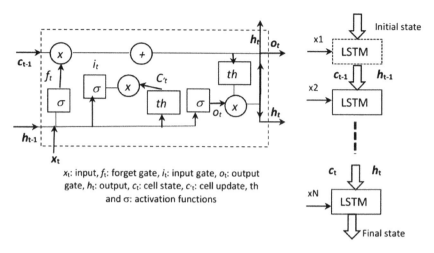

Fig. 13.5 Architecture of a LSTM neural network

these long-term dependencies. Figure 13.5 shows a simplified architecture of a LSTM neural network.

13.3 Photovoltaic Systems

PV systems consist mainly of a PV array, batteries, converters, inverters and protection devices. There are three main types of photovoltaic systems: (1) stand-alone, the batteries are always used (indispensable) and can be employed particularly in remote areas (isolated sites, such as Sahara). (2) hybrid PV systems, in this type more than one renewable (e.g., wind) or conventional (e.g., hydraulic) sources can be integrated. (3) grid-connected PV systems, these kind of systems are the most used for industrial and grid applications. In point of view capacity, PV systems can be classified as small scale (up to 10 kW) for domestic applications, medium scale for industries (less than 1 MW) and large-scale utility grid (up to 10 of MW).

13.3.1 Monitoring PV Systems

The main tasks of Monitoring PV systems (MPVSs) [11] include the monitoring of the meteorological and electrical data, in order to control and supervise the PV plant in real time. Figure 13.6 shows a simplified block diagram of a MPVS.

Advanced MSs should integrate the Internet of Things (IoT) technique, to check the performance and the system evolution in real time. Recently, efforts have

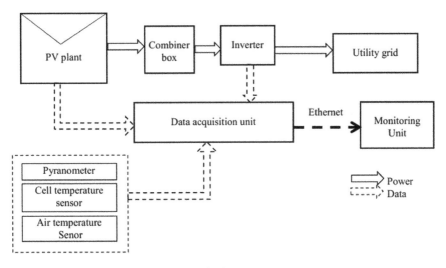

Fig. 13.6 Simplified schematic of a PV monitoring system

been focused on the integration of IoT in order to develop smart-monitoring systems for remote sensing of PV plants [12, 13]. As reported in [14], PIC or ARM Cotex-M3 Processor-based system can be used in large size plants with industrial aspects.

13.3.2 Faults in Photovoltaic Arrays

As reported in [4] faults in a PV array can be grouped into two main categories: permanent and temporally. Permanent faults are, for example, delamination, bubbles, yellowing, scratches and burnt cells. So, this category of faults can be cleared simply by replacing the faulty modules. Temporal faults are basically due to partial shading effects, dust accumulation (soiling), dirt on PV modules and snow that can be removed by operators without replacing the faulty PV module [15]. The most catastrophic faults are: arc, ground and line-to-line faults, and other faults that have a negative impact on the system are: hotspot, bypass and blocking diodes faults [16].

13.3.3 Faults Diagnosis Techniques

Several fault diagnosis methods have been proposed, and the main features that can characterize such method are: to detect the malfunctions quickly, input data (climatic and electrical data) and selectivity (ability to distinguish between different faults). They can be globally classified into two main categories:

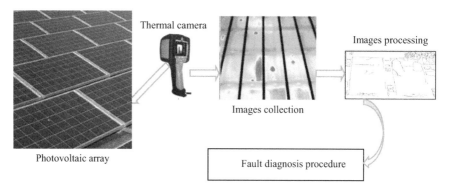

Photovoltaic array

Thermal camera

Images collection

Images processing

Fault diagnosis procedure

Fig. 13.7 Simplified diagram of fault diagnosis-based thermal image processing

(a) **Methods-Based on image processing**

These methods can be used for detecting most common defects on PV modules (e.g., discoloration, browning, surface soiling, hot spot, breaking and delamination). Drones are mainly used to localize and identify faults based on some advanced image processing algorithms, including DL. These methods are able identify and localize the faults, however, sophisticated equipment or instruments are needed, such as thermal cameras and drones. Figure 13.7 depicts a simplified diagram of a fault diagnosis method based on thermo-photography.

(b) **Electrical methods**

Electrical methods can be used for detecting and diagnosing faulty PV modules, strings and arrays including arc fault, grounding fault, diodes fault, and line to line fault. Some electrical-based fault diagnosis methods rely on some type of PV systems model to detect various types of faults. Electrical methods can be also classified globally into three groups:

- *Statistical and signal processing approaches*

Statistical and signal processing methods are mainly based on the analysis of the waveform signals; for example, time domain reflectometry, speared spectrum time domain reflectometry and earth capacitance measurement. These methods can be used to detect and localize faults; however, they are not able to identify all type of the faults. These techniques are not automatic and they are suitable for small-size PV plants.

- *Methods-based I-V characteristics analysis*

These methods are based on the analysing of the *I-V* curves of the PV arrays (See Fig. 13.8). A data-acquisition system is required in order to collect and store *I-V* curves, and then, based on the measured values of solar irradiance and solar cell temperature, simulated *I-V* curves can be plotted based on an accurate model (e.g., one diode model or other improved versions). Based on the measured and the

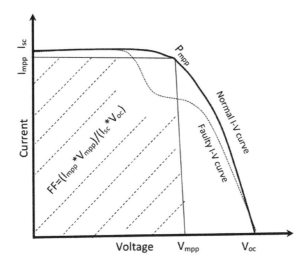

Fig. 13.8 *I-V* curves of a PV array

calculated maximum power, this method compares some points from the *I-V* curves in order to get an idea on the type of the faults (i.e., decease in Isc and Voc, variation on Vmpp and Impp, multiple local maximum power points). However, in some cases, these methods are not able to identify faults that have the same symptoms [4]. They can localize some type of faults only in very small-size PV plants, and it requires many sensors to localize faults at PV module level.

- *Methods-based artificial intelligence technique*

These methods include machine learning and recently deep learning. In these methods, MPVSs are indispensable to collect different data, such as meteorological and electrical data from PV systems. Figure 13.9 shows the principal working of these

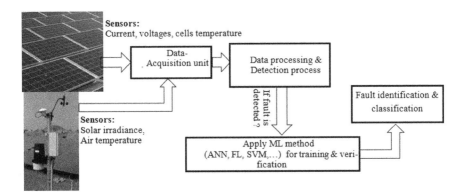

Fig. 13.9 Block diagram of fault diagnosis-based AI techniques

methods. These methods are able to classify and identify faults. They can be used to distinguish between faults that have the same symptoms, i.e., it can be got the same faulty I-V curves, but the origin of the fault is completely different. It should be noted that the identification accuracy depends mainly to the amount of the data, data quality and the used techniques [4].

13.4 Fault Diagnosis Based AI Techniques

This section is aiming at showing only fault diagnosing method-based AI techniques. Recently, many papers have been published on the application of AI in fault diagnosis of PV systems. Generally, these methods use as input some meteorological and electrical parameters or thermal images, in order to identify and classify the faults, so a database is needed to run these methods. They use also a large dataset of thermal or infrared images to identify faults in large-size PV plants. Table 13.1 lists some selected applications of AI techniques in fault diagnosis, detection and localization in PV systems.

With reference to the above Table 13.1, the following points can be highlighted:

- It can be clearly observed that a remarkable progress on the using of AI techniques in fault diagnosis of PV systems, from 2003 to 2019. Researchers are more and more attracted by machine learning and recently deep learning. It has been shown that supervised learning is the most widely used algorithm.
- Most advanced methods-based machine learning require a dataset of electrical parameters, such as measured currents and voltages, in some cases meteorological data are also needed (Such as solar irradiance, air temperature, and other parameters).
- The most investigated techniques in this field are neural networks and fuzzy logic, mainly used for fault classification in order to distinguish between faults that have the same signatures and classify type of faults. Overall, developed methods-based AI technique have been validated by simulation using MATLAB.
- The most available fault diagnosis techniques have been evaluated offline and only for small-size PV plants. In addition, most investigated faults are the ones appeared in a PV array (e.g., short-circuit, open-circuit and hot spot).
- Real-time implementation and cost of such technologies should be considered. To date, a very limited number of methods-based AI technique are verified at laboratories by developing a laboratory-scale prototype.
- The accuracy of the designed fault diagnosis-based AI depends mainly on the quality and the amount of the used data. Therefore, data should be carefully pre-treated and analysed. In addition, an appropriate learning algorithm should be also selected based on the solving problem (e.g., classification, identification or localisation).

Table 13.1 Reviewing papers on the fault detection, identification and localization of photovoltaic systems based on AI techniques

References	Year	Comments
[17]	2003	Expert system-based learning method is developed in order to diagnosis a PV system. The designed technique is used to diagnose the shading effect. The method was simulated and validated experimentally. In point of view complexity, we can say that the method is simple and effective but only in the case of shedding effect
[18]	2009	Fault diagnosis method-based neural network is developed, for some permanent faults. The method is applied for a small-scale grid-connected PV system, and it can be implemented easily with a medium cost
[19]	2010	An artificial neural network is used to build an intelligent fault diagnosis in a small-scale PV system. The proposed method can detect some temporally fault accurately with less time. The system is relatively easy to implement
[20]	2011	A kind of neural networks named Bayesian belief network is developed to detect temporally and permanent faults in a grid-connected PV system. The method was simulated for offline application. The method is focused more on data-collection. At this stage, the method is not able to diagnosis the fault. An expensive instrument is required to achieve this method
[21]	2011	A fuzzy logic approach was used to detect fault in a PV module. The method is relatively easy in point of view implementation, but it costs as the system needs more sensors to localize the faults
[22]	2011	An automatic detection fault-based fuzzy logic is developed and showed it ability to identify more than 90% of fault conditions in a PV array. The method can be used for temporally and permanent faults
[23]	2012	A fuzzy logic classifier is designed to detect an increase in series resistance (Rs) and verified experimentally. Results showed a good detection rate over a wide range of irradiance levels. But the technique needs more skill in point of view implementation
[24]	2014	Support vector machine (SVM) and k-nearest neighbour (k-NN) approaches are used to detect and classify short circuit in a PV array. The method was tested and simulated under MATLAB/Simulink. Only short circuit is investigated, but the method requires more knowledge in machine learning
[25]	2014	An extension neural network fault diagnosis method is used to identify whether the PV power generation system is operating normally or a fault has occurred. The method is relatively easy in point of view implementation. Validated and tested experimentally PV array. It can be used for both type of faults (permanent and temporal)
[26]	2015	Fuzzy logic is applied for evaluating the diagnostic rules. The designed method can be used for detecting partial shading, increased series-resistance losses and potential-induced degradation in a PV system. The main characteristic of the methods is fast and low computation. It can be used for both type of faults (permanent and temporal)

(continued)

Table 13.1 (continued)

References	Year	Comments
[27]	2015	An intelligent technique-based neural network was designed to identify faults in a PV system. The technique has been verified experimentally, and the highest probability of detection was 86%. It can be used for temporally fault detection
[28]	2015	A cost-effective algorithm-based a cascade fuzzy logic for detecting arc fault in a PV system is developed, the method is verified by simulation for a permanent fault
[29]	2016	A neural network-based model is designed to detect fault in a PV array. The method was tested for a temporal fault, which is shading effect. The method is relatively easy to implement and it was validated under MATLAB/Simulink for a PV array
[30]	2016	A fuzzy logic technique for fault detection in a PV array was developed. The designed algorithm is able to discriminate between the most frequently occurring PV module faults, such as increased series losses, bypass diode and blocking diode with good accuracy (90–98%)
[31]	2017	A method based on theoretical I-V curves analysis and FL classification system for fault detection in DC side of a 1.1 kWp grid-connected PV system was developed. The investigation fault is partial shading effect on PV modules. The method is relatively easy to be implemented and it was also verified experimentally
[32]	2017	An approach based on machine learning and *I-V* curves is developed to detect and classify permanent and temporally fault in a PV system. The method was validated under simulation-based MATLAB/Simulink using experimental data
[33]	2018	A multiclass fuzzy logic model based MATLAB/Simulink is designed to classify some permanent faults. The method was tested using an emulator system. The results showed that the model is able to classify the investigated faults with good accuracy
[34]	2018	A fuzzy logic and a radial basis function networks are used and compared in order to detect fault in PV systems. Both methods were used to detect and classify permanent and temporally faults
[35]	2018	A method based on C-means clustering fuzzy logic was designed to detect short-circuit, open circuit, partial shading and other faults. The method was tested and simulated offline
[36]	2019	Fuzzy logic approach is used to detect different hot spot type in a PV arrays. Extensive simulation and experimental-based tests have been carried out and the method showed good accuracy
[37]	2019	A fault diagnosis technique combines a long short-time memory (LSTM) network and softmax regression classifier is designed. The LSTM network is used to extract features, and the last one is used as input to a regression classifier. The model is used to test hot spot and line-to-line faults with good accuracy

(continued)

Table 13.1 (continued)

References	Year	Comments
[38]	2019	A deep conventional neural network is used to diagnose faults in PV array, the investigated faults are line-to-line and the current and the voltage are converted into electrical 2D graph to feed the DCNN. The method was used only for fault classification with good accuracy
[39]	2019	Machine learning and thermography was used to classify hot spot fault in PV modules. The method outperforms the classification approaches and provides good results. But, the method is relatively expensive due to the used equipment

13.5 Conclusions

In this chapter, the literature on fault diagnosis of PV systems using AI techniques (from 2003 to 2019) is briefly reviewed. It should be pointed that AI techniques have recently attracted many researchers working in this field, and many efforts have been deployed in order to design and implement new advanced methods for fault diagnosis of PV systems.

Accurately identification the cause or the nature of the fault, multiple faults, and the fault localization remains a challenging issue, particularly in large-size PV plants. Few attempts have been carried out using thermal images (collected by a unmanned aerial vehicle, e.g., drones) to localize defects in PV array. It should be noted that fault diagnosis accuracy using machine learning algorithms is relatively not scalable, particularly with a large size database. However, deep learning algorithms are the most appropriate in this case, due to their capability to manage a huge number of data. The application of AI techniques in fault diagnosis of PV systems, particularly deep learning, will continue to progress in the near future.

Acknowledgements The author would like to thank the Simons Foundation for financial support. A part of this work was carried out at the ICTP, Trieste, Italy.

References

1. Masson, G., Kaizuka, I.: Trends 2018 in photovoltaic applications, IEA-PVPS, Paris, France, Rep (2018). Available at: http://www.iea-pvps.org/fileadmin/dam/intranet/task1/
2. Daliento, S., Chouder, A., Guerriero, P., Pavan, A.M., Mellit, A., Moeini, R., Tricoli, P.: Monitoring, diagnosis, and power forecasting for photovoltaic fields: a review. Int. J. Photoenergy (2017). https://doi.org/10.1155/2017/1356851
3. Triki-Lahiani, A., Abdelghani, A.B.B., Slama-Belkhodja, I.: Fault detection and monitoring systems for photovoltaic installations: a review. Renew. Sustain. Energy Rev. **82**, 2680–2692 (2018)
4. Mellit, A., Tina, G.M., Kalogirou, S.A.: Fault detection and diagnosis methods for photovoltaic systems: a review. Renew. Sustain. Energy Rev. **91**, 1–17 (2018)

5. Mellit, A., Kalogirou, S.A.: Artificial intelligence techniques for photovoltaic applications: a review. Prog. Energy Combustion Sci. **34**(5), 574–632 (2008)
6. Russell, S.J., Norvig, P.: Artificial intelligence: a modern approach, 3rd edn. Prentice-Hall Inc, USA (2009)
7. Arthur, S.: Some studies in machine learning using the game of checkers. IBM J **3**, 211–229 (1959)
8. Alpaydin, E.: Machine learning: the new AI. MIT press, USA (2016)
9. LeCun, Y., Bengio, Y., Hinton, G.: Deep learning. Nature **521**(7553):436 (2015)
10. Goodfellow, I., Bengio, Y., Courville, A.: Deep learning. MIT Press, New York (2016)
11. Molina-García, A., Campelo, J.C., Blanc, S., Serrano, J.J., García-Sánchez, T., Bueso, M.C.: A decentralized wireless solution to monitor and diagnose PV solar module performance based on symmetrized-shifted gompertz functions. Sensors. **15**, 18459–18479 (2015)
12. Hamied, A., Boubidi, A., Rouibah, N., Chine, W., Mellit, A.: IoT-based smart photovoltaic arrays for remote sensing and fault identification. In: International conference in artificial intelligence in renewable energetic systems. Springer, Cham, pp. 478–486 (2019)
13. Hamied, A., Mellit, A., Zoulid, M.A., Birouk, R.: IoT-based experimental prototype for monitoring of photovoltaic arrays. In: 2018 International Conference on Applied Smart Systems (ICASS), IEEE, pp. 1–5 (2018)
14. Rahman, M.M., Selvaraj, J., Rahim, N.A., Hasanuzzaman, M.: Global modern monitoring systems for PV based power generation: a review. Renew. Sustain. Energy Rev. **82**, 4142–4158 (2018)
15. Haque, A., Bharath, K.V.S., Khan, M.A., Khan, I., Jaffery, Z.A.: Fault diagnosis of photovoltaic modules. Energy Sci Eng **7**(3), 622–644 (2019)
16. Köntges, M., Kurtz, S., Jahn, U., Berger, K., Kato, K., Friesen, T., et al.: Review of failures of photovoltaic modules. In: IEA PVPS Task, p. 13 (2014)
17. Yagi, Y., Kishi, H., Hagihara, R., Tanaka, T., Kozuma, S., Ishida, T., et al.: Diagnostic technology and an expert system for photovoltaic systems using the learning method. Solar Energy Mater Solar Cells **75**, 655–663 (2003)
18. Wu, Y., Lan, Q., Sun, Y.: Application of BP neural network fault diagnosis in solar photovoltaic system. In: IEEE international conference on mechatronics and automation, pp. 2581–2585 (2009)
19. Chao, K.-H., Chen, C.-T., Wang, M.-H., Wu, C.-F.: A novel fault diagnosis method based-on modified neural networks for photovoltaic systems. In Advances in swarm intelligence. Springer, Berlin, pp. 531–539 (2010)
20. Coleman, A., Zalewski, J.: Intelligent fault detection and diagnostics in solar plants. In: The 6th IEEE International Conference on Intelligent Data Acquisition and Advanced Computing Systems (IDAACS), pp. 948–953 (2011)
21. Cheng, Z., Zhong, D., Li, B., Liu, Y.: Research on fault detection of PV array based on data fusion and fuzzy mathematics. In: IEEE Asia-Pacific power and energy engineering conference, pp. 1–4 (2011)
22. Ducange, P., Fazzolari, M., Lazzerini, B., Marcelloni, F.: An intelligent system for detecting faults in photovoltaic fields. In: 11th IEEE international conference on intelligent systems design and applications, pp. 1341–1346 (2011)
23. Spataru, S., Sera, D., Kerekes, T., Teodorescu, R.: Detection of increased series losses in PV arrays using fuzzy inference systems. In: 38th IEEE photovoltaic specialists conference, pp. 464–469 (2012)
24. Rezgui, W., Mouss, L.H., Mouss, N.K., Mouss, M.D., Benbouzid, M.: A smart algorithm for the diagnosis of short-circuit faults in a photovoltaic generator. IEEE First Int Conf Green Energy ICGE **2014**, 139–143 (2014)
25. Chao, K.H., Chen, P.Y., Wang, M.H., Chen, C.T.: An intelligent fault detection method of a photovoltaic module array using wireless sensor networks. Int. J. Distrib. Sens. Netw. **10**(5), 540147 (2014)
26. Spataru, S., Sera, D., Kerekes, T., Teodorescu, R.: Diagnostic method for photovoltaic systems based on light I-V measurements. Sol. Energy **119**, 29–44 (2015)

27. Jones, C.B., Stein, J.S., Gonzalez, S., King, B.H.: Photovoltaic system fault detection and diagnostics using laterally primed adaptive resonance theory neural network. In: IEEE 42nd Photovoltaic Specialist Conference (PVSC), pp. 1–6 (2015)
28. Grichting, B., Goette, J., Jacomet, M.: Cascaded fuzzy logic based arc fault detection in photovoltaic applications. In: IEEE International Conference on Clean Electrical Power (ICCEP), pp. 178–183 (2015)
29. Mekki, H., Mellit, A., Salhi, H.: Artificial neural network-based modelling and fault detection of partial shaded photovoltaic modules. Simul Modelling Pract Theor **67**, 1–13 (2016)
30. Belaout, A., Krim, F., Mellit, A.: Neuro-fuzzy classifier for fault detection and classification in photovoltaic module. In: 8th IEEE International Conference on Modelling, Identification and Control (ICMIC), pp. 144–149 (2016)
31. Dhimish, M., Holmes, V., Mehrdadi, B., Dales, M., Mather, P.: Photovoltaic fault detection algorithm based on theoretical curves modelling and fuzzy classification system. Energy **140**, 276–290 (2017)
32. Chen, Z., Wu, L., Cheng, S., Lin, P., Wu, Y., Lin, W.: Intelligent fault diagnosis of photovoltaic arrays based on optimized kernel extreme learning machine and IV characteristics. Appl. Energy **204**, 912–931 (2017)
33. Belaout, A., Krim, F., Mellit, A., Talbi, B., Arabi, A.: Multiclass adaptive neuro-fuzzy classifier and feature selection techniques for photovoltaic array fault detection and classification. Renew Energy **127**, 548–558 (2018)
34. Dhimish, M., Holmes, V., Mehrdadi, B., Dales, M.: Comparing Mamdani Sugeno fuzzy logic and RBF ANN network for PV fault detection. Renew Energy **117**, 257–274 (2017)
35. Zhao, Q., Shao, S., Lu, L., Liu, X., Zhu, H.: A new PV array fault diagnosis method using fuzzy C-mean clustering and fuzzy membership algorithm. Energies **11**(1), 238 (2018)
36. Dhimish, M., Badran, G.: Photovoltaic hot-spots fault detection algorithm using fuzzy systems. IEEE Trans. Device Mater. Reliab. (2019). https://doi.org/10.1109/tdmr.2019.2944793
37. Appiah, A.Y., Zhang, X., Ayawli, B.B.K., Kyeremeh, F.: Long short-term memory networks based automatic feature extraction for photovoltaic array fault diagnosis. IEEE Access **7**, 30089–30101 (2019)
38. Lu, X., Lin, P., Cheng, S., Lin, Y., Chen, Z., Wu, L., Zheng, Q.: Fault diagnosis for photovoltaic array based on convolutional neural network and electrical time series graph. Energy Convers. Manage. **196**, 950–965 (2019)
39. Kurukuru, V.B., Haque, A., Khan, M.A.: Fault classification for photovoltaic modules using thermography and image processing. In: IEEE industry applications society annual meeting, pp. 1–6 (2019)